Towards Sustainability

Nichole Hugo

# Towards Sustainability

## Exploring Challenges Beyond the Environment

Springer

Nichole Hugo
Sustainability
Eastern Illinois University
Charleston, IL, USA

ISBN 978-3-032-29338-1  ISBN 978-3-032-29339-8 (eBook)
https://doi.org/10.1007/978-3-032-29339-8



This Springer imprint is published by the registered company Springer Nature Switzerland AG
The registered company address is: Gewerbestrasse 11, 6330 Cham, Switzerland

# Preface

## Sustainability Discourse and the Overlooked Dimensions of Social and Economic Issues

Sustainability is a topic that can feel overwhelming. While the slow pace of generating substantial change, coupled with numerous barriers, contributes to this vexation, it is not the main source of this frustration. The disagreements about how to move forward with a positive impact on our communities could also be seen as trying. However, a critical concern is the lack of a unified understanding of sustainability and its comprehensive nature. If we are not talking about the same thing, how are we able to work on solving these issues together?

Sustainability is often explained through three main pillars: environmental, economic, and social. Yet much of the discussion, especially in academic literature, focuses heavily on environmental issues, while the social and economic dimensions are often overlooked. This imbalance and the consequences of ignoring social and economic factors are explored in this book.

Historically, environmental protection has dominated sustainability discussions, particularly as climate change, biodiversity loss, and resource depletion have become pressing global concerns. The United Nations' 1987 Brundtland Report, which first defined sustainability, highlighted the interconnectedness of environmental, social, and economic dimensions (World Commission on Environment and Development, 1987). Despite this, studies show that most sustainability literature focuses on environmental issues. Researchers like Smith et al. (2008) note that social and economic dimensions are often treated as secondary, while Geissdoerfer et al. (2017) show that topics such as carbon emissions, renewable energy, and waste reduction continue to dominate sustainability frameworks. This leaves less room for discussions on the social and economic implications of sustainable development.

While environmental sustainability is extensively discussed, social sustainability is often underrepresented or represented as a niche topic in sustainability. Social sustainability, which includes equitable distribution of resources, social justice, and improving quality of life for marginalized groups, is frequently underrepresented (Barron et al., 2023). However, social dimensions are frequently sidelined in sustainability literature, even though social sustainability is crucial for achieving long-term environmental and economic goals. Boström (2012) argues that the emphasis on environmental goals often ignores how sustainability impacts communities, culture, and human rights. The failure to integrate social issues such as labor conditions, human rights, and health impacts results in a narrow understanding of sustainability. Yet, many sustainability strategies fail to acknowledge the necessary integration of these goals, leading to policies that are, in practice, unsustainable due to their inability to address social justice.

The economic pillar of sustainability is another dimension that is often marginalized in discussions centered predominantly on environmental sustainability. Economic sustainability concerns the creation of economic systems that are not only efficient, but also equitable and resilient. It includes considerations of wealth distribution, equitable access to resources, and fair-trade practices (Elkington, 1998). However, economic considerations like long-term economic resilience, equitable economic systems, and the role of sustainable business models are often glossed over in favor of discussions focused on carbon emissions, pollution control, and ecosystem services. Without integrating economic development into sustainability frameworks, policies risk neglecting the financial feasibility and long-term scalability of sustainability solutions.

Beyond academics sounding the call for a more holistic look at sustainability, the United Nations Sustainable Development Goals broadened the scope of their goals from the Millennium Goals, and National Geographic attempted to reach the public with new sustainable terminology to emphasize these other pillars.

The United Nations began with 8 sustainability goals in 2000, called the Millennium Development Goals (United Nations, 2000). These goals focused on environmental, economic, and social factors to be complete between 2000-2015. This helped shed more light on sustainability initiatives and assisted with driving policy in a way that was understandable and measurable for multiple areas of sustainability.

With the conclusion of the Millennium Development Goals in 2015, the Sustainable Development Goals were introduced, expanding the goals to 17 to be reached by 2030. These encompassed and looked more closely at issues previously explored with the Millennium Goals, but further expanded into areas of social justice and quality of life. This expanded set of goals delved into areas completely ignored under the umbrella of sustainability, such as cyberbullying under SDG 16-Peace, Justice and Effective Institutions. While this global initiative gives a reference and justification for including these forgotten topics in the sustainability discussion, they still tend to be overlooked or placed within silos instead of being included in sustainable practices.

National Geographic tried to tackle this problem in the field of tourism where societal, cultural, and economic issues abound. National Geographic has strived to educate the public about sustainability by aiming to inspire a global audience to engage with sustainability issues in an approachable and meaningful way. The organization's efforts extend beyond environmental protection to include the integration of cultural and economic dimensions in sustainable practices. One of its most notable contributions to this field is the creation of the term "geotourism," which is a concept that broadens the scope of sustainable tourism by emphasizing the preservation of both natural and cultural heritage. By coining the term, National Geographic introduced a new way of thinking about how tourism can support sustainability in diverse and interconnected ways. Geotourism, as defined by National Geographic, is "tourism that sustains or enhances the geographical character of a place—its environment, culture, aesthetics, heritage, and the well-being of its residents" (National Geographic, 2002). While this is essentially the same concept as ecotourism, the idea of forming a new term to highlight and emphasize areas outside of the environment pushed the public to rethink sustainable tourism. The idea of ecotourism emerged from the realization that conventional tourism often damages local ecosystems, communities, and cultures. Unlike traditional forms of tourism, which can contribute to environmental degradation and cultural erosion, ecotourism encourages travelers to engage in activities that preserve the very qualities that make a destination unique. However, this term has been hijacked by marketing companies as a buzzword and improperly used to lure tourists to areas, regardless of if sustainable social, economic, and environmental practices are in place. Reintroducing the idea with a new term was a good idea, but the impact has been limited.

With all of these attempts to broaden the scope of sustainability, why has it been so difficult to move away from a laser focus on environmental protection? Several factors contribute to the predominance of environmental concerns in sustainability literature and discussion. One key factor is the urgency and visibility of environmental crises, such as climate change and biodiversity loss, which are often perceived as immediate global threats. These pressing challenges demand urgent action, which leads to a focus on environmental solutions. Additionally, environmental issues are often easier to measure and quantify compared to social or economic dimensions, which are more complex and context-dependent (UNEP, 2015).

Furthermore, political and economic interests may play a role in shaping the dominance of environmental sustainability discussions. As global markets become increasingly driven by eco-friendly technologies and renewable energy industries, there is significant investment in promoting environmental narratives (Celestin et al., 2024). This investment often leads to an overemphasis on environmental factors, especially in business and policy sectors.

However, the marginalization of social and economic sustainability can have serious implications. Failing to consider the social impacts of sustainability initiatives can lead to exacerbating inequality, creating new forms of exclusion, or undermining local communities. For example, policies promoting renewable energy technologies, such as large-scale hydropower or biofuels, can result in the displacement of communities or harm to local economies if social impacts are not adequately considered. Similarly, neglecting economic sustainability can hinder the scalability of environmental policies. Without considering the economic feasibility of sustainability strategies, policies may fail to secure long-term investment or may overlook important considerations such as job creation, economic equity, and local economic development.

While environmental sustainability continues to dominate academic and policy discussions, the social and economic dimensions of sustainability remain insufficiently explored. This imbalance has implications for both theory and practice, as a narrow focus on environmental issues risks neglecting essential factors like social justice, economic equity, and long-term resilience. As the global community seeks to address the complex and interconnected challenges of the 21st century, future sustainability research must broaden its scope to fully integrate the environmental, economic, and social pillars. This more inclusive approach will provide a more holistic understanding of sustainability and promote more effective, equitable solutions for global challenges. This book serves as a call to reshape the way we have, en masse, traditionally researched sustainability topics and enacted policy. Viewing sustainability without a holistic framework will lead to the continuation of the environmental problems researchers and practitioners have been trying to fix. Without economic planning and societal changes, the fragile environment will continue to be jeopardized.

Charleston, IL, USA Nichole Hugo

## References


Barron, P., Cord, L., Cuesta, J., Espinoza, S., Larson, G., & Woolcock, M. (2023). *Social sustainability in development: Meeting the challenges of the 21st century*. World Bank.

Boström, M. (2012). A missing pillar? Challenges in theorizing and practicing social sustainability. *Sustainability Science Practice and Policy, 8*(1), 3–14.

Celestin, M., Vasuki, M., Sujatha, S., & Kumar, A. D. (2024). Implementing green technologies to reduce environmental impact: Economic and competitive advantages of eco-friendly practices. *International Journal of Scientific Research and Modern Education*, 9(2), 33–39.

Elkington, J. (1998). *Cannibals with forks: The Triple Bottom Line of 21st century business*. Capstone.

Geissdoerfer, M., Savaget, P., Bocken, N. M., & Hultink, E.J. (2017). The Circular Economy: A new sustainability paradigm? *Journal of Cleaner Production, 143*, 757–768.

National Geographic. (2002). Geotourism. https://www.nationalgeographic.com/maps/topic/geotourism#:~:text=Geotourism,well%2Dbeing%20of%20its%20residents

Smith, A., Stirling, A., & Berkhout, F. (2008). The governance of sustainable socio-technical transitions. *Research Policy*, *38*(4), 495–510.

UNEP. (2015). *Global environment outlook: Regional assessments. Cambridge University Press*. United Nations Environment Programme.

United Nations. (2000). *Millennium development goals and beyond 2015*. https://www.un.org/millenniumgoals/

United Nations. (2015). Sustainable development: The 17 goals. https://sdgs.un.org/goals

World Commission on Environment and Development. (1987). *Our common future*. Oxford University Press.

# Contents

# About the Author

**Dr. Nichole Hugo** is a Professor in Hospitality and Tourism and the Graduate Coordinator of Sustainability at Eastern Illinois University. Her research centers on sustainable development in tourism, with particular emphasis on how destinations, especially in developing regions, balance economic vitality, community well-being, and environmental stewardship. She has examined topics such as tourism's role in poverty alleviation, destination image and resilience, cultural and community engagement in festivals and events, and the adoption of sustainable practices across lodging and tourism sectors. Her work also extends to environmental planning and community facing sustainability initiatives, including land use planning, climate adaptation efforts in small and medium-sized communities, and strategies for enhancing tourism's contribution to local development. Across her publications, technical reports, and book chapters, Dr. Hugo integrates marketing, community development, and sustainability perspectives to better understand how tourism and related industries can create more equitable and environmentally responsible outcomes.

# Chapter 1
# What Is Sustainability?

## 1.1 What Does It Mean to Be Sustainable?

Have you ever heard the word "sustainable" or "sustainability" and wondered what it actually means? You're not alone. The term *sustainability* gets thrown around so often, and sometimes incorrectly, that it can be confusing to pin down its real meaning.

Many people think sustainability is just about protecting the environment. While the environment is a big part of it, focusing only on that can actually lead to unsustainable outcomes in other areas.

So, what *is* sustainability?

The term was first officially defined in the **Brundtland Report** as "meeting the needs of the present without compromising the ability of future generations to meet their own needs" (UNED, 1987, p. 43). While this is an important starting point to define sustainability, it doesn't really tell us much. This report was published in 1987 by Gro Harlem Brundtland through the United Nations and it played a major role in getting the world to think about sustainable development (UNED, 1987). The Brundtland report is widely cited as the foundation for later international agreements and frameworks on sustainable development (Borowy, 2013; Hajian & Kashani, 2021).

The United Nations later built on this report by creating global targets. First came the **Millennium Development Goals (MDGs)** from 2000 to 2015, which focused mostly on reducing extreme poverty, improving education, and health (United Nations, 2015; World Health Organization, 2015). These were later followed by the **Sustainable Development Goals (SDGs)** which are a set of 17 goals adopted by nearly every country in the world in 2015. These goals are meant to be achieved by 2030 and cover sustainability issues such as ending hunger, fighting climate change, as well as promoting peace and social justice (UNDP, n.d.)


N. Hugo, *Towards Sustainability*, https://doi.org/10.1007/978-3-032-29339-8_1

## 1.2 The Three Pillars of Sustainability

While the Brundtland definition gives us a general idea of what sustainability means, it's pretty vague and hard to measure. That's why sustainability is often explained as a balance between three main pillars of development:

- **Economic Sustainability**: This means supporting jobs, businesses, and financial systems in a way that helps people make a living now *and* in the future. It's about making sure resources are used efficiently, workers are treated fairly, and communities can build long-term prosperity.
- **Socio-cultural Sustainability**: This focuses on people, traditions, communities, education, health, equity, and the protection of cultural identities and traditions. A sustainable society supports people's basic needs and ensures everyone has the chance to thrive.
- **Environmental Sustainability**: It's about protecting ecosystems, conserving natural resources, and reducing pollution and waste. Environmental sustainability ensures that our planet remains livable by focusing specifically on maintaining the health and resilience of natural systems so they can continue providing clean air, clean water, fertile soil, and other essential ecological services that we need to survive.

**Sustainability** means finding a balance between these three pillars. It's about making decisions that boost the economy, support communities, and protect the environment—*all at the same time*. That balance helps societies create smart policies and grow in ways to help future generations to improve or sustain their wellbeing.

One way to make sustainable efforts more effective is by creating **linkages** between the three pillars. That means looking at how economic, social, and environmental goals can support each other, rather than trying to improve each one in isolation. The idea is that if you improve one you can hopefully strengthen one of the other pillars as well. (We'll dive deeper into what "linkages" are in later chapters.)

While this sounds like a great idea it is actually very tricky to do. When you improve one or two pillars, it often causes problems for the third. These are called **tradeoffs**. For example building a school might help improve education for children in the area, but might take the place of where a business could have been built or may need trees to be removed for development. Trade-offs are hard to avoid, so the key is deciding whether the positive outcomes are worth the negative ones.

## 1.3 Finding the Balance (and Why It's Not Easy)

In theory, the idea of balancing all three pillars sounds great. Who wouldn't want a society that's fair, prosperous, and environmentally responsible? However, since improving one area can sometimes make another one worse we need to think about responsible planning and measuring impacts. Let's look at an example where it appears that protecting the environment is beneficial, but can have negative impacts to local wildlife and communities.

**Take A Closer Look: Hydropower** A hydropower dam is typically seen as a great way to bring clean energy to an area, which can reduce carbon emission, produce less pollution than traditional forms of energy production, and create jobs for locals in the area. However, multiple researchers have noted the downside to the environment and livelihoods for the people in nearby communities. For example, research by Castro-Diaz et al. (2023) shows that these projects often come with serious downsides for local communities that includes displacement of locals, disruption to wildlife and negative impacts to local businesses. Dams may provide new roads, power lines, or housing, but they usually take away far more in terms of land, livelihoods, and community well-being. Families are frequently forced to move, losing farmland, fishing grounds, and cultural traditions tied to the rivers. When people are relocated, their social networks are broken up, trust within communities weakens, and health and education often suffer. Promises of jobs rarely work out because most of the work requires skilled engineers, leaving local people with few long-term benefits.

For instance, the Nam Theun 2 Dam in Laos displaced over 6200 people on the Nakai Plateau and has had severe downstream effects on more than 100,000 people, many of whom lost access to farming and fishing and became dependent on limited or inadequate cash compensation (BankTrack, 2025; International Rivers, 2010). Critics argue that resettled households often received poor-quality land, insufficient livelihood support, and inadequate compensation, leaving many with unsustainable incomes (Radio Free Asia, 2019). In Brazil's Amazon, major dams such as the Belo Monte Dam have displaced tens of thousands of people but failed to deliver equitable energy or economic benefits. The Belo Monte project has been associated with declines in rural employment, disrupted livelihoods, and increased reliance on off-grid energy sources; some resettled communities continued to rely on diesel generators, and the social costs have included reduced social capital, worsening health, and increased inequality (Bro et al., 2018; Mayer et al., 2021, 2022). Moreover, research on the Jirau and Santo Antônio dams along the Madeira River shows that resettlement led to declines in traditional fishing and floodplain agriculture, with communities pressured to adopt market-based livelihoods that eroded their well-being (Roquetti et al., 2024). Overall, while hydropower can look like a win for clean energy on paper, this research shows it can bring damage to those nearby.

We'll explore more of these examples later in the book. For now, it's important to remember that sustainability is not just about focusing on the environment. It's about thinking about development holistically and trying to build systems to support people and the planet. We are going to take a closer look at each of the three pillars of sustainability so we can have a better idea of how we can have them benefit one another.

## 1.4 Why Is It Important to Study Sustainability?

If global organizations and businesses are already focused on sustainability, *why do I need to learn about it?* Even though sustainability is a topic that is being discussed on a global scale, real change depends on individuals, communities and leaders working together. The global population is expected to reach 9.7 billion by 2050 (United Nations, 2019), which puts huge pressure on our resources, economies, and societies. While some industries are making progress, many are still lagging behind in addressing key sustainability issues like fair wages, community well-being, and responsible production. Over 700 million people still live in extreme poverty, surviving on less than $2.15 per day (World Bank, 2023). This shows how urgently we need economic sustainability, particularly as our population grows.

Without widespread education on sustainability, we risk repeating the same harmful patterns that has led to crucial action being needed. Studying sustainability helps you understand how interconnected everything is. When industries ignore fair labor practices or when cities grow without planning for social equity, both people and the planet suffer. By learning about sustainability, you're better equipped to challenge the status quo, make smarter decisions, and improve conditions for the future.

## 1.5 What Industries Should Be Focusing on Sustainability?

Most industries can contribute to being sustainable and make a difference by strengthening communities, supporting livelihoods, and boosting economic resilience. Sustainability has become increasingly relevant across sectors as industries recognize that long term success depends on responsible resource use and positive social impact. Below is a brief overview of industries that have strong ties to sustainability:

**Political Science** Governments shape the rules that define sustainability in practice. The World Bank in 2024 launched a 10-year "Scaling Up Energy Efficiency in Europe and Central Asia" program (E3) that provides $1.46 billion in financing to upgrade public buildings, modernize industry, and retrofit district heating systems (World Bank, 2024). This initiative is expected not only to reduce $CO_2$ emissions by millions of tons but also to create jobs in construction, manufacturing, and energy services, thereby contributing to economic resilience and growth. Such policies demonstrate how government-led sustainability strategies can generate meaningful social returns in addition to environmental gains. In addition, policy can assist with social justice, education, and economic initiatives.

***A Closer Look: Brazil's Bolsa Família Program*** Brazil's Bolsa Família, a conditional cash transfer program, reduced extreme poverty by more than 50% between 2003 and 2014. By tying financial support to children's school attendance and healthcare visits, the program strengthened both economic security and social well-being across generations (World Bank, 2014).

**Fashion and Merchandising** The fashion industry contributes roughly 10% of global carbon emissions and 20% of industrial wastewater (UNEP, 2018). Tackling sustainability means reducing water use, shifting to circular business models, moving away from fast-fashion, and ensuring fair labor practices (Gautam, 2024).

***A Closer Look: Fair Trade Cotton in India*** Fair trade initiatives in India have improved wages and working conditions for cotton farmers. As India produces nearly 25% of the world's cotton, roughly 5.8 million farmers are impacted by fair trade practices (Fair Trade India, 2022).

**Business** Businesses that adopt sustainable practices, such as fair-trade sourcing, worker protections, and waste reduction, tend to gain competitive advantage. As the public becomes more mindful about their products and how they are made, more companies are motivated to make sustainable changes to meet the demands of their consumers.

***A Closer Look: Ben & Jerry's Social Mission*** Ben & Jerry's webpage explains how its business centers around values like fair trade and living wages. The company pays its workers above average wages and supports local farmers with fair prices, demonstrating how businesses can grow profitably while prioritizing worker welfare and community development (Ben & Jerry's, 2022; Fair Trade America, 2025).

**Tourism** Tourism is responsible for around 8% of the world's greenhouse gas emissions (Lenzen et al., 2018). If unchecked, tourism's energy use could rise 154% and emissions 131% by 2050 (Srivastava, 2025). Sustainable tourism shifts the focus to protecting communities and giving back to local economies. Since mass amounts of people visiting an area can erode the local culture and environment, understanding how to attract and manage tourists throughout their stay is important to minimize damage to the destination, particularly for fragile areas.

***A Closer Look: Homestay Tourism in Nepal*** In Nepal community-based homestay tourism allows families in rural areas to host travelers. The income helps support education for children, medical care, and the preservation of traditional crafts and food practices, strengthening cultural pride while diversifying rural economies (Woli, 2022). Tourists not only have a unique experience, but they are able to learn and broaden their understanding of the local culture in a way mass operated tour businesses cannot replicate.

**Schools and Universities** Schools and universities educate the next generation and can also model sustainability through equity and access initiatives. Many educational institutions benefit directly from World Bank-supported programs that improve infrastructure and inclusion (World Bank, 2024).

***A Closer Look: University of Cape Town's Student Equity Programs*** The University of Cape Town runs programs to expand access for students from disadvantaged backgrounds. By combining financial aid, mentorship, and cultural inclusion efforts, the university has made higher education more socially sustainable, bridging gaps in equity and opportunity across South Africa (University of Cape Town).

**Healthcare** Healthcare systems produce about 4.4% of global greenhouse gas emissions, which is roughly equivalent to 514 coal powered pants (Health Care Without Harm, 2019). In the U.S. alone, hospitals accounted for 8.5% of the nation's total emissions in 2018 (Eckelman & Sherman, 2018). Proper healthcare management is also important for maintaining strong communities, as having healthy individuals means less resources are needed to improve overall health and they are more likely to be able to be productive members of society. Hospitals that adopt sustainable models also improve affordability and community well-being. As sustainability also has a large focus on improving quality of life and longevity, healthcare plays a vital role in sustainable initiatives.

***A Closer Look: Rwanda's Community Health Worker Program*** Rwanda's national health insurance program, *Mutuelle de Santé*, covers 90% of the population (Think Global Health, 2021). It has helped lower the amount people pay out of pocket for healthcare, from 28% down to 12% of total health costs, and increased how often people use health services to about 1.8 visits per person each year. To make the system fairer, Rwanda also uses a program called *Ubudehe*, which was originally created to support vulnerable people after the 1994 genocide and now helps communities categorize households based on income, assets, and ability to work, so that the poorest families receive full subsidies and don't have to pay for insurance (Makaka et al., 2012).

## 1.6 Historical Events

Understanding how different industries contribute to sustainability is just one part of the big picture. To truly grasp how far we've come, it's important to look at the historical milestones that have shaped sustainability efforts worldwide. Over the past few decades, international summits, treaties, and reports have laid the groundwork for today's sustainability practices. These events influenced policies, industries, and even the way we think about sustainable development from a holistic perspective. The timeline below highlights key moments that pushed sustainability to a central focus in global decision-making.

**1987—Brundtland Report (Our Common Future)** The Brundtland Report introduced the modern definition of sustainable development as meeting present needs without compromising future generations' ability to meet theirs. It highlighted the links between economic growth, environmental protection, and social equity, shaping global sustainability discourse (World Commission on Environment and Development, 1987).

**1992 Earth Summit (Rio de Janeiro, Brazil)** The Earth Summit brought together world leaders to address global environmental and development challenges. It produced key frameworks including *Agenda 21*, a global action plan for sustainability, and the *Rio Declaration on Environment and Development*, which set guiding principles for balancing growth and environmental stewardship (United Nations, 1992).

**1997 Kyoto Protocol** The Kyoto Protocol was the first international treaty to establish binding targets for reducing greenhouse gas emissions. Participating countries agreed to cut emissions during the 2008–2012 period, marking a significant step in global climate governance despite challenges in implementation (UNFCCC, 1997).

**2002 World Summit on Sustainable Development (Johannesburg, South Africa)** This summit assessed progress since the 1992 Earth Summit and highlighted the need to integrate sustainability with poverty reduction. It emphasized access to clean water, energy, and sustainable consumption, shifting focus from commitments toward practical implementation (United Nations, 2002).

**2012 Rio+20 Conference (Rio de Janeiro, Brazil)** Rio+20 revisited the global sustainability agenda, leading to the establishment of the High-Level Political Forum on Sustainable Development. It also called for the creation of the Sustainable Development Goals (SDGs), setting the stage for a unified global framework (United Nations, 2012).

**2015 UN Adopts the 2030 Agenda for Sustainable Development** In September, the United Nations launched the 17 Sustainable Development Goals (SDGs), a universal framework for tackling poverty, inequality, and environmental challenges by 2030. The goals built on past commitments but provided a clearer, measurable roadmap for sustainable progress (United Nations, 2015).

**2015 Paris Agreement (COP21)** In December, nations adopted the Paris Agreement, a historic accord to limit global warming to well below 2 °C (3.6 °F), with an ambition to cap it at 1.5 °C (2.7 °F). The agreement required all countries to submit climate action plans, shifting responsibility beyond industrialized nations alone (UNFCCC, 2015).

**2023 SDG Summit (New York)** At the halfway point toward the 2030 deadline, the SDG Summit reviewed progress and found the world falling behind on many targets. Leaders called for urgent acceleration, especially in addressing climate change, education, and inequality, to keep the goals within reach (United Nations, 2023).

## 1.7 Wrap-Up

Sustainability is more than just a buzzword for protecting the environment, as it is a critical framework for protecting and balancing economic, social and cultural, and environmental assets. Sustainability touches multiple aspects of our lives and careers and our actions can provide meaningful change when we have individuals, communities and businesses work together on these initiatives. However, as we have seen, the path to sustainability is complex and finding the right balance between the three pillars requires thoughtful decision making and a willingness to embrace both challenges and tradeoffs.

The historical milestones, like the Brundtland Report and the adoption of the Sustainable Development Goals, show us that we have made significant progress, but there is still much work to be done. These events remind us that sustainability is not just a concept discussed at the global level, but a guiding idea that shapes the decisions we make in our own communities. As industries continue to grow and populations expand, the need to consider the three pillars becomes even more important. When economic growth ignores social well-being or when environmental protection does not consider the needs of local people, the results can create new problems instead of solving existing ones. This is why a balanced and informed approach is essential.

Moreover, sustainability requires ongoing learning, adaptability, and innovation. As new technologies emerge, markets evolve, and social dynamics shift, the strategies we use to support sustainable development must also adapt. Collaboration across sectors, countries, and communities becomes critical to solving complex challenges like climate change, poverty, and resource scarcity. By embracing sustainable practices in both small-scale daily actions and large-scale industrial or policy decisions, we can create systems that are more resilient, inclusive, and equitable.

As individuals, industries and governments work toward a sustainable future, it is important to remember that sustainable development is a shared responsibility and the choices we make today will determine the quality of life for future generations. By understanding how the three pillars connect and how they influence one another, we can make better decisions in our daily lives and in our careers. In the chapters ahead, we will take a closer look at each pillar and explore how different industries apply sustainability in practical and meaningful ways.

## References


BankTrack. (2025). *Nam Theun 2 dam: Social and human rights impacts.* https://www.banktrack.org/project/nam_theun_2_dam

Ben & Jerry's. (2022). *Our values.* https://www.benjerry.com/values

Borowy, I. (2013). *Defining sustainable development for our common future: A history of the world commission on environment and development (Brundtland commission).* Routledge.

Bro, A. S., Calvi, M. F., & Moran, E. F. (2018). The Belo Monte hydroelectric dam and its impact on rural employment and food production. *Sustainability, 10*(5), 1592. https://doi.org/10.3390/su10051592

Castro-Diaz, L., García, M. A., Villamayor-Tomas, S., & Lopez, M. C. (2023). Impacts of hydropower development on locals' livelihoods in the global south. *World Development, 169.* https://www.sciencedirect.com/science/article/pii/S0305750X23001031

Eckelman, M. J., & Sherman, J. (2018). Estimated global disease burden from U.S. healthcare sector greenhouse gas emissions. *American Journal of Public Health, 108*(S2), 120–122. https://pubmed.ncbi.nlm.nih.gov/29072942/

Fair Trade America. (2025). *Helping Ben & Jerry's provide living incomes.* https://www.fairtrade.net/us-en/for-business/benefits-of-being-certified/business-success-stories/helping-ben%2D%2D-jerry-s-provide-living-incomes.html

Fair Trade India. (2022). *Cotton.* https://www.fairtradeindia.org/cotton

Gautam, R. (2024). The narrative of circular economy and sustainability: A critical analysis of fashion industry. *Circular Economy and Sustainability, 4*(4), 3183–3213.

Hajian, M., & Kashani, S. J. (2021). Evolution of the concept of sustainability. From Brundtland report to sustainable development goals. In *Sustainable Resource Management* (pp. 1–24). Elsevier.

Health Care Without Harm. (2019). *Health care's climate footprint: How the health sector contributes to the global climate crisis and opportunities for action.* https://global.noharm.org/sites/default/files/documents-files/5961/HealthCaresClimateFootprint_092319.pdf

International Rivers. (2010, April 5). *Laos' Nam Theun 2 Dam starts operation.* River Resource Hub. https://riverresourcehub.org/resources/laos-nam-theun-2-dam-starts-operation-3478

Lenzen, M., Sun, Y. Y., Faturay, F., Ting, Y. P., Geschke, A., & Malik, A. (2018). The carbon footprint of global tourism. *Nature Climate Change, 8*(6), 522–528. https://doi.org/10.1038/s41558-018-0141-x

Makaka, A., Breen, S., & Binagwaho, A. (2012). Universal health coverage in Rwanda: A report of innovations to increase enrolment in community-based health insurance. *The Lancet, 380.* https://www.thelancet.com/journals/a/article/PIIS01406736%2813%2960293-7/fulltext

Mayer, A., Johansen, I. C., López, M. C., & Moran, E. F. (2022). Changes in social capital associated with the construction of the Belo Monte dam: Comparing a resettled and a host community. *Global Environmental Change, 74,* 102487. https://doi.org/10.1016/j.gloenvcha.2022.102487

Mayer, A., Lopez, M. C., Johansen, I. C., & Moran, E. F. (2021). Hydropower, social capital, community impacts, and self-rated health: Evidence from Amazonian dam-affected communities. *Rural Sociology, 86*(4), 829–864. https://doi.org/10.1111/ruso.12355

Radio Free Asia. (2019, April 11). *Lao villagers displaced by Nam Theun 2 Dam say they still need help.* https://www.rfa.org/english/news/laos/displaced-04112019122447.html

Roquetti, D. R., Athayde, S., Silva-Lugo, J., & Moretto, E. M. (2024). Amazon communities displaced by hydroelectric dams: Implications for environmental changes and householdś livelihood. *Global Environmental Change, 89*, 102933.

Srivastava, U. (2025, July 31). *Overtourism: A growing challenge for people, places and policy.* The Analysis. https://theanalysis.org.in/overtourism-a-growing-challenge-for-people-places-and-policy

Think Global Health. 2021. *Rwanda's health-care success holds lessons for others.* https://www.thinkglobalhealth.org/article/rwandas-health-care-success-holds-lessons-others.

UNDP. (n.d.). *Sustainable development goals.* United Nations Development Programme. https://www.undp.org/sustainable-development-goals

UNED (1987). *Our common future*. Oxford University Press. United Nations World Commission on Environment and Development. https://sdgs.un.org/sites/default/files/documents/5987our-common-future.pdf

UNEP. (2018). *Putting the brakes on fast fashion.* United Nations Environment Programme https://www.unep.org/news-and-stories/story/putting-brakes-fast-fashion

UNFCCC (1997). *The Kyoto protocol.* United Nations Framework Convention on Climate Change. https://unfccc.int/process-and-meetings/the-kyoto-protocol

UNFCCC. (2015). *The Paris agreement.* United Nations Framework Convention on Climate Change. https://unfccc.int/process-and-meetings/the-paris-agreement

United Nations. (1992). *United Nations conference on environment and development (UNCED).* https://www.un.org/en/conferences/environment/rio1992

United Nations. (2002). *World summit on sustainable development.* https://sustainabledevelopment.un.org/milesstones/wssd

United Nations. (2012). *Rio+20—United Nations conference on sustainable development.* https://sustainabledevelopment.un.org/index.php?menu=35&nr=23&page=view&type=400

United Nations. (2015). *Transforming our world: The 2030 agenda for sustainable development.* https://sdgs.un.org/2030agenda

United Nations. (2019). *World population prospects 2019: Highlights.* Department of Economic and Social Affairs, Population Division. https://population.un.org/wpp/assets/Files/WPP2019_Highlights.pdf

United Nations. (2023). *2023 SDG summit: United Nations.* https://www.un.org/en/conferences/SDGSummit2023

Woli, L. (2022). Impact of homestays on socio-economic opportunities of the local community. *KMC Journal, 4*(2), 212–223.

World Bank. (2014, March 22). *How to reduce poverty: A new lesson from Brazil for the world?* World Bank. https://www.worldbank.org/en/news/feature/2014/03/22/mundo-sin-pobreza-leccion-brasil-mundo-bolsa-familia

World Bank. (2023). *Poverty headcount ratio at $2.15 a day.* https://data.worldbank.org/indicator/SI.POV.DDAY

World Bank. (2024, June 27). *World Bank unveils major initiative to scale up energy efficiency in Europe and Central Asia.* https://www.worldbank.org/en/news/press-release/2024/06/25/world-bank-unveils-major-initiative-to-scale-up-energy-efficiency-in-europe-and-central-asia

World Health Organization. (2015). *Millennium development goals (MDGs).* https://www.who.int/news-room/fact-sheets/detail/millennium-development-goals-%28mdgs%29

# Chapter 2
# Environmental Impacts

## 2.1 The Environmental Pillar of Sustainability

Of the three pillars of sustainability (environmental, socio-cultural, and economic), the environmental pillar tends to be he one most people immediately associate with the idea of sustainability. It is also the easiest to understand because it deals directly with the natural systems that keep our planet livable.

When ecosystems like forests, wetlands, or coral reefs get destroyed, the ripple effects directly impact us. Food systems get less stable, water supplies dry up, and even the air we breathe gets dirtier. About 1 million species are at risk of extinction because of habitat loss, pollution, and overuse of natural resources (IPBES, 2019; World Bank). That's not just sad for wildlife, as it is also threatening to the balance that keeps humans alive. This means it is critically important to protect biodiversity, reduce pollution, manage forests and oceans wisely, and mitigate the effects of climate change. Without a healthy environment, neither societies nor economies can thrive.

In addition to these direct effects, environmental degradation can intensify conflicts over scarce resources and increase economic instability. Water shortages, soil degradation, and declining fish stocks often create competition between communities, industries, and even nations, putting social cohesion and economic development at risk. Protecting ecosystems also supports essential regulatory functions, such as flood control, disease regulation, and climate moderation, which are often taken for granted until disrupted. Furthermore, intact ecosystems provide cultural and recreational benefits, supporting mental health, education, and tourism-based economies. By investing in environmental protection, societies not only preserve natural systems but also build resilience to shocks and stresses, ensuring that communities can adapt to change and maintain livelihoods over time.



N. Hugo, *Towards Sustainability*, https://doi.org/10.1007/978-3-032-29339-8_2

## 2.2 Why Environmental Protection Matters

Environmental degradation directly harms human health in many ways. Air pollution is the fourth-leading cause of premature death worldwide, responsible for more than 7 million deaths each year (World Bank, 2023). In many developing countries, polluted water sources contribute to diarrheal diseases that kill hundreds of thousands of children annually. By reducing pollution and protecting ecosystems, sustainability efforts improve quality of life and reduce preventable health risks.

India faces some of the worst air pollution levels in the world, with New Delhi often recording air quality index (AQI) levels far above safe limits. It was found that in 2022 more than 1.7 million people in India died due to exposure to fine particulate matter (Lancet, 2025). While economic growth, industrial activity, and vehicle use have created jobs and raised living standards, they have also worsened public health. This demonstrates the trade-off between industrial development and environmental sustainability.

Efforts like promoting electric vehicles and expanding renewable energy capacity are being introduced, but enforcement remains a challenge. These initiatives also have their own downsides, with issues such as battery creation and disposal being harmful to the environment. With a growing population with growing needs, finding ways to meet demand through effective use of resources and limiting waste is imperative.

Loss of flora (plants) and fauna (animals) disrupts ecosystems by removing species that play important roles in pollination, soil health, and food webs. As habitats shrink or degrade, many plants and animals struggle to survive, leading to declines in biodiversity. These changes can reduce ecosystem stability, making natural systems less resilient to stresses like climate change. Over time, the loss of species can also affect human communities that depend on healthy ecosystems for resources, livelihoods, and cultural values.

Rainforest destruction leads to the loss of countless plant species that hold potential medicinal properties, many of which have not yet been studied. As these ecosystems shrink, valuable traditional knowledge about healing plants can disappear along with the communities that use and protect them. Rainforests also play a major role in regulating the Earth's atmosphere by absorbing carbon dioxide and producing oxygen, often being described as the "lungs of the planet." Their continued decline threatens both global biodiversity and important natural processes that support life on Earth.

The list of negative environmental impacts around the world could fill a whole textbook, but below are some examples of some of the negative impacts on the environment around the world to help illustrate the destruction being caused. These include:

- **Deforestation:** Approximately 10 million hectares of forest are lost each year (Earth.org, 2025; FAO, 2021). That's roughly the size of Switzerland disappearing annually.

- **Plastic Waste in Oceans:** Around 8 million metric tons (with high estimates at 12.7 and low at 4.8 million tons) of plastic enter the oceans from land in coastal countries each year (Jambeck et al., 2015; Jambeck Research Group, 2025). To visualize, that's like placing five grocery bags of plastic for every foot of coastline (Jambeck Research Group, 2025).
- **Municipal Waste:** The world generates about 2.01 billion tons of municipal solid waste per year (Kaza et al., 2018), which is like filling roughly 800,000 Olympic-sized swimming pools (assuming one pool holds ~2500 tons).
- **Food Waste:** Roughly one-third of all food produced for people, which is about 1.3 billion tons, is lost or wasted every year (FAO, 2011). That amount could fill over 500,000 Boeing 747 s (if each holds ~250 tons).

These environmental challenges highlight how closely human well-being is tied to the health of natural systems. When air, water, soils, and ecosystems deteriorate, the impacts extend beyond wildlife and landscapes, they also affect economic stability and the industries people rely on. Understanding these connections is essential for addressing environmental problems in ways that support both development and long-term resilience. This leads directly to how major global industries depend on, and are increasingly shaped by, the condition of the environment.

## 2.3 Links Between Environment and Industry

Many of the world's most important industries depend on a healthy environment. Agriculture relies on fertile soils, predictable weather, and clean water. Yet, the Food and Agriculture Organization (FAO) warns that 33% of soils worldwide are already degraded due to erosion, deforestation, and chemical overuse (FAO, 2020). Similarly, fisheries support the livelihoods of more than 200 million people globally, but over one-third of fish stocks are overexploited (FAO, 2022; World Bank, 2022).

Another industry heavily impacted by environmental degradation is healthcare, where pollution driven illnesses place significant strain on medical systems and the workforce. Poor air quality is linked to elevated rates of cardiovascular disease, stroke, chronic respiratory conditions, and lung cancer, contributing to millions of premature deaths each year (WHO, 2023). These environmental health burdens increase healthcare expenditures, lengthen wait times, and reduce productivity as workers experience more illness related absences.

The technology sector is also directly affected by environmental conditions. Semiconductor manufacturing requires large quantities of ultra-pure water for rinsing and cooling processes, with individual fabrication facilities using millions of gallons per day (NIST, 2024). As global chip demand increases, water use in semiconductor production is expected to rise substantially, intensifying pressure on water-scarce regions (IDTechEx, 2023). Data centers also rely on significant water inputs for cooling and indirectly contribute to water stress through their high electricity consumption, much of which still comes from thermoelectric power plants that require large water withdrawals (EESI, 2023).

Other industries, like tourism, are also highly dependent on the environment. For some developing countries, the pristine environment is a crucial asset to draw people to the destination and improve their economies. Coral reefs, beaches, and national parks attract millions of visitors annually, generating billions in revenue. However, coral reefs are dying at alarming rates due to ocean warming and pollution, which threatens both biodiversity and tourism economies. In some small island developing states, tourism accounts for over 30% of GDP (UNWTO, 2022). If natural attractions disappear, so too will a vital source of income.

The Caribbean relies heavily on reef-based tourism, which generates more than $6 billion annually (UNEP, 2021). However, rising sea temperatures, overfishing, and pollution are causing widespread coral bleaching. The loss of reefs not only threatens marine biodiversity but also undermines coastal protection and the tourism-dependent economies of small island states. The case in the Caribbean illustrates the deep link between natural ecosystems and industry sustainability.

## 2.4 Resource Management and Long-Term Growth

One of the central ideas behind the environmental pillar of sustainability is that natural resources must be managed responsibly to ensure their continued availability for future generations. This principle emphasizes the need for a balance between human development and ecological preservation. When societies overexploit natural systems, such as forests, freshwater sources, and fertile soils, they not only endanger ecosystems but also jeopardize the foundations of their own long-term economic stability. Forests, for example, provide a wide range of essential ecosystem services, as they supply timber and other raw materials, regulate local and global water cycles, stabilize soils, and act as significant carbon sinks that mitigate climate change. However, these benefits are being rapidly diminished as deforestation continues at alarming rates across the globe.

However, forested areas equivalent to the size of Portugal are degraded or go through desertification every year, largely driven by agricultural expansion and unsustainable land-use practices (World Bank, 2023). As farming practices use intensive tillage and chemical use, soil erosion and contamination contribute to a decline in land quality. This large-scale deforestation reduces biodiversity, increases greenhouse gas emissions, and weakens the planet's ability to regulate climate patterns. Beyond forests, similar patterns of overuse threaten other natural systems, such as overfishing in marine environments and groundwater depletion in agricultural regions, which further highlight the urgent need for better resource management strategies.

Sustainable practices like crop rotation, reforestation, and water conservation are essential tools for reversing these negative trends. Crop rotation helps maintain soil fertility and prevents pest buildup, reforestation restores degraded landscapes and captures atmospheric carbon, and water conservation ensures that freshwater supplies remain available even in times of drought. When governments, industries, and communities invest in such practices, they not only preserve the environment but also create the conditions for steady, long-term economic growth. Sustainable resource management promotes resilience, allowing societies to adapt to environmental changes and maintain productivity over time. Responsible resource management thus lies at the heart of both environmental integrity and long-term growth.

## 2.5 Understanding Climate Change

Perhaps the most urgent environmental issue of our time is climate change. **Climate change** refers to long-term alterations in temperature, precipitation, and weather patterns due to human activity, primarily the burning of fossil fuels. These impacts include more severe weather events, such as flooding, drought, and more frequent and intense storms.

**Global warming** is one aspect of climate change, but it specifically refers to the rising average surface temperatures of the Earth caused by greenhouse gas emissions (National Geographic, 2025). In other words, global warming is the heating, while climate change is the broader set of consequences that result from it. The United Nations reports that the average global temperature is already 1.2 °C (2.2 °F) higher than pre-industrial levels and recent years have been the warmest on record (United Nations, 2023).

As this warming leads to extreme weather events, such as stronger hurricanes, prolonged droughts, and catastrophic flooding, it can be more difficult for societies to recover from its impact. The World Bank estimates that by 2030, climate change could push over 100 million people into extreme poverty, particularly in Sub-Saharan Africa and South Asia (World Bank, 2015). Issues with climate change are not only an environmental concern, as economic and socio-cultural aspects will also be impacted as resources become limited due to increases in catastrophic weather. As people relocate to avoid adverse weather areas, changes in cultural and societal norms will shift. The ability for people to find new jobs may be difficult and resources will continue to become more expensive as they become more limited. This is why understanding climate change is an important part of tackling sustainability challenges for all three pillars.

## 2.6 Greenhouse Gases Explained

Greenhouse gases are gases in the atmosphere that trap heat, making the Earth warmer. The main greenhouse gases include carbon dioxide ($CO_2$), methane ($CH_4$), nitrous oxide ($N_2O$), and fluorinated gases. Human activities such as burning coal, oil, and gas for energy release vast amounts of these gases. As a result, carbon dioxide concentrations are now 50% higher than pre-industrial levels, the highest in at least 2 million years (UNEP, 2022).

Reducing greenhouse gas emissions is essential to slowing climate change. Strategies include shifting to renewable energy, improving energy efficiency, and protecting forests that naturally absorb $CO_2$. The World Bank has invested heavily in renewable energy projects, with over 35% of its climate-related financing in 2022 directed toward clean energy transitions (World Bank, 2023).

The importance of understanding this can be seen in the Amazon rainforest. The Amazon rainforest was long considered one of Earth's most important carbon sinks because its vast forests stored large amounts of carbon dioxide (UNEP, 2021). However, recent research covering the period from 2010 to 2018 shows that the region has become a net source of carbon to the atmosphere, mainly due to fire emissions (Basso et al., 2023). Fires, often linked to clearing land for cattle ranching and soybean farming, release large quantities of carbon and sharply reduce the forest's capacity to reabsorb $CO_2$. In addition, droughts and rising temperatures weaken forest resilience and slow regrowth (Basso et al., 2023). As a result the Amazon no longer reliably offsets global emissions, but instead contributes to them. This shift matters because losing the Amazon's role as a carbon sink accelerates climate change, undermines biodiversity, harms the ecosystem services the forest provides, and heightens the tension between short-term economic gains from agriculture and long-term global climate stability.

## 2.7 Environmental Justice and Global Inequality

One of the challenges in sustainability is the unequal way environmental damage and climate change affect different parts of the world. Many developed countries built their economies through early industrialization, which relied heavily on practices such as burning coal, clearing forests, and polluting waterways. These activities were understood differently at the time, and the long-term environmental consequences were not yet clear. Today, however, developing countries are encouraged to follow cleaner development paths, even as many are still working to meet basic needs and expand essential services.

This creates differences not only in economic capacity but also in how environmental impacts are felt. For example, Africa contributes less than 4% of global carbon emissions, yet it is among the regions most vulnerable to climate impacts like droughts and floods (UNEP, 2023). Roughly 389 million people in Sub-Saharan Africa still lack access to electricity, and climate change makes progress even harder (World Bank, 2022). As weather patterns become less predictable, challenges such as food insecurity, displacement, and infrastructure strain become more difficult to manage.

These realities highlight the complexity of global sustainability efforts. Regions that contributed relatively little to past emissions may face significant environmental pressures simply because of geography, climate patterns, or economic structure.

Many communities rely on stable natural conditions for farming, water access, and daily livelihoods, so even small environmental shifts can have large social and economic effects. These changes may require families to adapt their work, relocate, or adjust long-standing traditions tied to land and seasons.

At the same time, developing countries often face difficult decisions when trying to balance growth with sustainability. Cleaner technologies and infrastructure improvements can be costly, require new skills, or take years to implement. While there is widespread agreement on the importance of sustainable development, the pace and pathways may look different depending on each country's resources, priorities, and immediate needs. This can create a sense of tension between long-term environmental goals and short-term development pressures.

A balanced approach to addressing these challenges involves recognizing the different starting points countries have. Meaningful progress depends on cooperation, knowledge sharing, and support that helps all countries pursue development in a way that strengthens both their economies and their resilience to climate impacts. This kind of collaboration allows sustainability to evolve as a shared effort rather than a burden shouldered unequally.

## 2.8 Wrap-Up

Protecting the environment is not just about saving animals or forests, as it involves safeguarding human health, cultural traditions, and economic futures. The environmental pillar of sustainability reminds us that without a healthy planet, social and economic systems will collapse. International organizations like the UN, UNEP, and the World Bank emphasize that solutions must be both ambitious and fair, with the goal that all countries, especially the most vulnerable, have the resources and support they need to adapt and thrive. However, this will require continuous involvement and support from individuals, communities, and world leaders.

Environmental change is both a global issue and one that affects people's daily lives in immediate, tangible ways. We see this in the degradation of air and water quality, which directly contributes to millions of premature deaths each year. It is evident in the strain placed on healthcare systems already stretched by pollution-related illnesses. It is present in the disruption of ecosystems that provide essential services, such as pollination, clean water, and fertile soils, that we depend on for food security and stable livelihoods. Environmental decline is not a distant or abstract concern, as it is already influencing how societies function, how economies grow, and how safely and healthily people can live.

The examples of soil degradation, collapsing fish stocks, deforestation, and plastic pollution illustrate how human activity has pushed natural systems beyond their limits. At the same time, the chapter shows how closely connected these systems are to major industries. Agriculture, fisheries, tourism, and technology all rely on stable and functioning ecosystems. When these systems deteriorate, industries face increased operating costs, reduced productivity, and long-term uncertainty.

Climate change intensifies all of these pressures. Rising temperatures, extreme weather events, and shifting precipitation patterns create new social and economic challenges, especially for regions with fewer resources to adapt. From droughts that reduce crop yields to floods that destroy infrastructure and displace communities, climate change multiplies existing vulnerabilities.

Ultimately, achieving environmental sustainability requires collective responsibility. Governments must create effective policies, industries must adopt cleaner technologies and more responsible practices, and individuals must make choices that reduce waste and conserve resources. None of these actions alone are enough, but together they form the basis for meaningful environmental protection. The path forward will require innovation, collaboration, and a commitment to balancing human needs with the limits of the natural world. By recognizing the interdependence of environmental, social, and economic systems, we can work toward a future where development does not come at the expense of the planet, but instead supports the well-being of all people for generations to come.

## References

Basso, L. S., Wilson, C., Chipperfield, M. P., Tejada, G., Cassol, H. L. G., Arai, E., Williams, M., Smallman, T. L., Peters, W., Naus, S., Miller, J. B., & Gloor, M. (2023). Atmospheric $CO_2$ inversion reveals the Amazon as a minor carbon source caused by fire emissions, with forest uptake offsetting about half of these emissions. *Atmospheric Chemistry and Physics, 23*, 9685–9723. https://doi.org/10.5194/acp-23-9685-2023

Earth.org. (2025). *10 shocking statistics about deforestation*. https://earth.org/statistics-deforestation/

Environmental and Energy Study Institute. (2023). *Data centers and water consumption*. https://www.eesi.org/articles/view/data-centers-and-water-consumption

FAO. (2011). *Global food losses and food waste: Extent, causes and prevention*. Food and Agriculture Organization. https://www.fao.org/4/mb060e/mb060e00.htm

FAO. (2020). *State of the world's soil resources*. Food and Agriculture Organization of the United Nations. https://openknowledge.fao.org/server/api/core/bitstreams/6ec24d75-19bd-4f1f-b1c5-5becf50d0871/content

FAO. (2021). *State of the world's forests 2020*. Food and Agriculture Organization. https://www.fao.org/4/mb060e/mb060e00.htm

FAO. (2022). *The state of world fisheries and aquaculture 2022: Towards blue transformation*. Food and Agriculture Organization of the United Nations. doi:https://doi.org/10.4060/cc0461en.

IDTechEx. (2023). *Water usage in semiconductor manufacturing to double by 2035*. https://www.idtechex.com/en/research-article/water-usage-in-semiconductor-manufacturing-to-double-by-2035/32746

IPBES. (2019). Global assessment report on biodiversity and ecosystem services of the intergovernmental science-policy platform on biodiversity and ecosystem services. In E. S. Brondízio, J. Settele, S. Díaz, & H. T. Ngo (Eds.), *Intergovernmental science-policy platform on biodiversity and ecosystem services*. doi:https://doi.org/10.5281/zenodo.3831673.

Jambeck, J. R., Geyer, R., Wilcox, C., Siegler, T. R., Perryman, M., Andrady, A., Narayan, R., & Law, K. L. (2015). Plastic waste inputs from land into the ocean. *Science, 347*(6223), 768–771. https://doi.org/10.1126/science.1260352

Jambeck Research Group. (2025). *Plastic input into the ocean–The 8 million ton estimate*. https://jambeck.engr.uga.edu/plastic-input-into-the-ocean-release-day

Kaza, S., Yao, L. C., Bhada-Tata, P., & Van Woerden, F. (2018). *What a waste 2.0: A global snapshot of solid waste management to 2050.* World Bank. https://documents.worldbank.org/en/publication/documents-reports/documentdetail/697271544470229584

Lancet. (2025). *The 2025 report of the Lancet countdown on health and climate change.* https://www.thelancet.com/journals/lancet/article/PIIS0140-6736(25)01919-1/abstract

National Geographic. (2025). *Global warming.* https://education.nationalgeographic.org/resource/global-warming/

National Institute of Standards and Technology. (2024). *Water quality and the semiconductor industry.* https://www.nist.gov/chips/chips-incentives-funding-opportunities/environmental-division/water-quality-and-semiconductor

UNEP. (2021). *Making peace with nature: A scientific blueprint to tackle the climate, biodiversity and pollution emergencies.* United Nations Environment Programme. https://digitallibrary.un.org/record/3901238

UNEP. (2022). *Emissions gap report 2022: The closing window.* United Nations Environment Programme. doi:https://doi.org/10.18356/9789210023993.

UNEP. (2023). *Africa's adaptation gap report.* United Nations Environment Programme. https://www.unep.org/resources/report/africas-adaptation-gap

UNWTO. (2022). *Tourism in small island developing states.* United Nations World Tourism Organization. https://www.unwto.org/small-island-developing-states

WHO. (2023). *Ambient outdoor air quality and health.* World Health Organization. https://www.who.int/news-room/fact-sheets/detail/ambient-(outdoor)-air-quality-and-health

World Bank. (2015). *Shock waves: Managing the impacts of climate change on poverty.* https://documents1.worldbank.org/curated/en/260011486755946625/pdf/Shock-waves-managing--the-impacts-of-climate-change-on-poverty.pdf

World Bank. (2022). *The World Bank and Oceans: The blue economy for sustainable development.* World Bank. https://www.worldbank.org/en/topic/oceans

World Bank. (2023). *Sustainable land management and forest conservation.* https://www.worldbank.org/en/topic/forests

# Chapter 3
# Socio-Cultural Impacts

## 3.1 The Socio-Cultural Pillar of Sustainability

As we learned in the previous chapter, learning about environmental factors like energy, forests, or reducing pollution is very important. However, we can't stop thinking about sustainability there. The socio-cultural pillar is equally important as it's all about giving people the tools and opportunities they need to improve their lives through education, social justice, and health. It's about nurturing resilient, inclusive communities, respecting cultural heritage, and fostering understanding across differences. When communities have equitable access to knowledge, justice, and health services, they are stronger, more resilient, and better equipped to face challenges (United Nations, 2025b; World Bank, 2025).

The issue is that this is a harder concept to understand in general, let alone remember when it comes to sustainability. So let's think about it in some other ways. When you turn on the news, you can hear some information about the weather and natural disasters that we can directly link to the environmental pillar of sustainability. However, we tend to hear more about politics and policy, crime and wars, debates about human rights, conflicts or negotiations between countries, and societal issues. This is why it is so important to remember this pillar, because it won't help if we can protect the environment if we can't learn how to get along with our neighbors in our communities and across the world.

Another reason this pillar is a challenge to learn about is because it can be called different things and cover a wide range of topics, as it deals with society and culture. Some researchers and organizations call it the *social pillar*, focusing on people, equity, and community well-being. Others use the term *cultural pillar*, emphasizing traditions, heritage, and identity. More often, it's referred to as the ***socio-cultural pillar***, which combines the two ideas of how society and culture shape human resilience and cooperation. Occasionally, you may even see it referred to as the *societal*


N. Hugo, *Towards Sustainability*, https://doi.org/10.1007/978-3-032-29339-8_3

*pillar*, a broader framing that highlights systems, governance, and collective organization.

Each of these terms points to a slightly different emphasis:

- **Social pillar:** highlights fairness, justice, inclusion, access to education, healthcare, and protection of rights.
- **Cultural pillar:** focuses on traditions, heritage, identity, diversity, and how these elements build belonging and resilience.
- **Socio-cultural pillar:** combines the social and cultural pillars, showing how these values are intertwined.
- **Societal pillar:** looks at institutions, governance, and the broader organization of communities and nations.

These overlapping terms sometimes make the concept seem abstract, but they all circle around the idea that sustainability is not only about protecting the planet and growing economies, but also about creating strong, fair, and cohesive societies that can thrive over time.

In this book, we will discuss some of these different terms, but we will usually refer to it as socio-cultural. We are also going to get into a little overlap with this pillar and the one that focuses on the economic impacts. For example, we will talk about aspects of improving quality of life through initiatives like fair pay and safe work environments, which ties directly to economic impacts. However, for the introductory purposes of this book, the socio-cultural impact has more of a focus on improving the lives of people, while the economic is focused more on resources. Let's dive deeper into the various aspects of human life and development to get a better understanding of the areas we need to consider within the socio-cultural pillar.

## 3.2 Education

Education is one of the central drivers of social sustainability. In 2023, 272 million children and youth were out of school, despite improvements in access to schooling at multiple levels (UNESCO, 2024c; United Nations, 2025a). Furthermore, even among children who attend school, more than half (roughly 600 million) are not reaching minimum proficiency levels in reading and math (UNESCO, 2017; UNICEF, 2025). This gap is especially stark in low and middle income countries, where it is estimated that 70% of 10-year-olds cannot demonstrate understanding of simple written text (World Bank, 2022).

Quality education builds critical thinking, expands opportunities, and strengthens communities. It enables young people to participate meaningfully in social, cultural, and economic life, equipping them with skills to address challenges such as poverty, inequality, and environmental degradation. However, achieving this requires sustained investment in inclusive classrooms, trained teachers, and safe learning environments, particularly for marginalized groups.

For instance, refugee children face systemic barriers. While 91% of children globally attend primary school, only 61% of refugee children do, and the figure drops to just 23% for refugee adolescents in secondary school (UNHCR, 2017). Addressing these gaps is essential for building social equity, empowering communities, and fostering long-term resilience across societies. Ensuring equitable access to education lays the foundation for sustainable development, connecting knowledge acquisition to future economic and socio-cultural growth.

## 3.3 Social Justice and Equity

**Social justice**, which includes ensuring fairness, inclusion, and rights for all, is essential to the socio-cultural pillar. Around 2 billion people in low and middle income countries still lack adequate social protection, which includes nearly three out of every four people in the poorest nations (World Bank, 2025). Expanding social safety nets not only helps people survive shocks such as illness or natural disasters, but also builds resilience and strengthens trust in institutions. Social justice involves more than protection, as it encompasses equal access to legal systems, participation in governance, and protection of cultural heritage.

The World Bank's Global Program on Justice and Rule of Law supports judicial reforms, access to legal aid, and community dispute resolution, which is key for protecting rights and heritage (World Bank, 2025). Gender equality is also central. Under the World Bank's Gender Strategy 2024–2030, efforts are underway to connect 300 million women to broadband, extend social protection to 250 million, and provide capital access to 80 million women-led businesses by 2030 (World Bank, 2024a, 2024b).

These initiatives show that addressing inequality and providing opportunities for marginalized groups strengthens communities' ability to engage in economic, social, and cultural development. When social justice is effectively implemented, people are empowered to participate fully in their societies, creating stronger, fairer, and more resilient communities capable of supporting sustainable growth.

## 3.4 Health and Community Resilience

Healthcare strengthens resilience, reduces poverty, and sustains cultural practices. When people have consistent access to healthcare, they are better able to participate in school, work, and community activities, which contributes to stronger and more equitable economic systems. Preventive care reduces the frequency and severity of illness, which lowers household medical costs and keeps families from falling deeper into poverty during health emergencies. Reliable healthcare systems also help communities recover more quickly from crises such as disease outbreaks or natural disasters by improving preparedness and response.

Health services also support cultural sustainability. Many communities maintain traditional healing knowledge, caregiving practices, and shared health rituals that influence how families understand wellness and resilience. When modern healthcare works alongside these traditions instead of replacing them, cultural identity is strengthened. This creates trust between providers and communities, encourages people to seek timely care, and preserves cultural practices that reinforce social cohesion.

Rwanda's Community Health Worker Program is a prime example. Since the 2000s, over 45,000 workers have been trained to provide rural healthcare. This approach reduced child mortality by more than 60% between 2000 and 2015, while healthcare coverage expanded to over 90% of the population (Binagwaho & Scott, 2015; World Bank, 2025). The program not only improved health outcomes but also created jobs and built trust at the community level. By bringing healthcare directly to rural populations, the program increased accessibility, particularly for women and children, who are often most vulnerable to preventable illnesses.

Community health workers serve as bridges between formal health systems and local populations, providing education on hygiene, nutrition, and disease prevention. These services help reduce the burden of disease while strengthening community resilience and social cohesion. Health improvements support economic participation and educational attainment, creating ripple effects that benefit families and communities. In addition, culturally sensitive approaches embedded in community health programs help preserve traditional health knowledge while integrating modern medical practices. This combination enhances trust, reinforces local identity, and ensures that healthcare contributes to sustainable socio-cultural development.

## 3.5 Cultural Heritage and Community Identity

**Cultural heritage**, which includes traditions, sacred sites, languages, and customs, anchors social identity. It is what ties people to their past and gives them a sense of belonging in the present. When sacred sites are destroyed, historic languages are lost, or cultural knowledge is neglected, entire communities lose resilience. Protecting culture is about preserving the living systems of knowledge, values, and practices that shape how societies understand themselves and interact with others.

One of the leading global organizations in this effort is UNESCO, which stands for the United Nations Educational, Scientific and Cultural Organization. Founded in 1945, UNESCO's mission is to build peace through international cooperation in education, science, culture, and communication. In the context of sustainability, UNESCO plays a critical role in safeguarding cultural diversity, advancing inclusive education, and ensuring that both tangible and intangible heritage are protected for future generations.

UNESCO's work on cultural heritage spans many areas:

- **World Heritage Sites:** UNESCO designates and monitors over 1100 cultural and natural sites worldwide, from the Great Wall of China to the historic city of Petra in Jordan. These sites are recognized for their "outstanding universal value" and protected under international agreements (UNESCO, 2024a).
- **Intangible Cultural Heritage:** Beyond physical sites, UNESCO also safeguards traditions like storytelling, music, rituals, and craftsmanship. For example, practices such as Japanese washoku (traditional cuisine) and the oral traditions of the Bedouin have been inscribed on UNESCO's Representative List of the Intangible Cultural Heritage of Humanity (UNESCO, 2024b).
- **Inclusive Education and Cultural Diversity:** UNESCO emphasizes that cultural diversity and inclusive education are inseparable. Protecting heritage must go hand-in-hand with teaching future generations about their own traditions as well as those of others. This creates tolerance, respect, and intercultural dialogue, which are essential elements of sustainable societies (UNESCO, 2025).

An example of their work can be seen in the tourism industry. Tourism is critical to Jordan's economy, but unmanaged tourism once endangered Petra, a fragile, yet popular attraction. Tourism has been a major driver of economic opportunity in the Petra region by creating jobs, supporting local businesses, and providing income that strengthens community resilience even during periods of regional instability. Tourism funding and international support has allowed for the advancement of conservation projects, expanded community programs, and funding went toward new visitor infrastructure that protects Petra's heritage. Initiatives such as the Nabatean Cultural Village, community markets, and training programs have helped diversify local livelihoods, reduce informal practices in the park, and ensure that the benefits of tourism directly support surrounding communities (UNESCO, 2024a).

Cultural heritage is not simply about the past. It is a living resource that, when protected, supports identity, strengthens community bonds, and contributes to sustainable development. UNESCO's global efforts highlight the need for policies that integrate heritage protection into education and social development.

## 3.6 Building Partnership Across Cultures

Learning from different cultures strengthens problem-solving and fosters innovation. When communities share perspectives and approaches, they can create solutions that are more creative, inclusive, and effective. For example, Indigenous communities often bring generations of ecological knowledge to natural resource management, while formal institutions contribute technical expertise and funding. When these forms of knowledge and authority are combined, societies can tackle complex challenges like climate change, public health, or conflict resolution in ways that are both locally relevant and globally effective.

Development projects that combine traditional leadership, civil society, and formal institutions tend to be more successful because they build trust and reflect the realities of the people most affected. The World Bank's Justice for the Poor initiative is a good example of this approach. This organization works with communities that use both customary (local/traditional) and formal (state-based) legal systems, helping to bridge the gap between the two. By recognizing the legitimacy of customary law while ensuring alignment with human rights standards, this approach enhances inclusivity and cohesion (World Bank, 2024a, 2024b).

In Indonesia, the World Bank's Justice for the Poor program collaborated with local communities in rural villages where disputes were traditionally resolved by elders or religious leaders. These customary systems had long been trusted, but they sometimes excluded women, youth, or minority groups from participating. The program introduced participatory processes that ensured broader community involvement while still respecting traditional practices.

One initiative involved land disputes, which were frequent due to overlapping claims between villagers and commercial developers. Rather than relying solely on the formal court system, which was expensive and often inaccessible, the program supported hybrid dispute-resolution models. These models combined the legitimacy of customary processes with legal safeguards to protect marginalized voices.

The outcomes were significant. Communities reported higher satisfaction with dispute outcomes, women gained more opportunities to have their concerns heard, and conflicts were resolved more quickly than through formal courts. Importantly, by strengthening trust between citizens and institutions, the program also reduced the likelihood of disputes escalating into violence (World Bank, 2024a, 2024b).

## 3.7 Wrap-Up

The socio-cultural pillar of sustainability is the backbone of thriving, equitable, and resilient societies. It reminds us that sustainability is not only about protecting the planet's ecosystems or ensuring economic prosperity, but also about empowering people and communities to live with dignity, opportunity, and purpose. When individuals have equitable access to education, justice, healthcare, and social participation, they become active contributors to sustainable development rather than passive recipients of aid. This empowerment strengthens societies from the inside out, creating communities that can withstand social, economic, and environmental challenges.

At its core, the socio-cultural pillar highlights the interdependence between people and the systems that shape their lives. A society cannot be sustainable if large portions of its population are denied access to basic education, if justice is not applied equally, or if cultural identities are erased or ignored. When people feel excluded, disempowered, or disconnected from their cultural roots, the social fabric weakens and cooperation, innovation, and resilience are harder to achieve. By contrast, when communities are inclusive and respectful of diversity, they can harness a broader range of knowledge, traditions, and perspectives to solve shared problems.

Global organizations such as the United Nations, UNESCO, and the World Bank have shown that social and cultural investment leads to tangible progress. Expanding educational opportunities builds human capital and opens doors to economic growth. Ensuring access to justice and equality under the law strengthens trust in institutions. Investing in healthcare protects lives while reinforcing community cohesion. And protecting cultural heritage, whether through language preservation, art, or historical sites, fosters identity, belonging, and continuity across generations. These investments are not only moral imperatives; they are practical strategies for achieving sustainable and inclusive development.

Moreover, the socio-cultural pillar encourages societies to view diversity as a strength rather than a source of division. In an increasingly interconnected world, cultural understanding and cooperation across borders are essential for addressing issues like climate change, migration, and global inequality. Initiatives that promote intercultural dialogue, gender equality, and participatory governance help reduce conflict and build mutual trust. Communities that embrace these values are better equipped to innovate, collaborate, and adapt in a rapidly changing global landscape.

In many ways, the socio-cultural pillar is what connects the environmental and economic dimensions of sustainability. Healthy societies are more capable of protecting their natural environments and using resources wisely, while inclusive economies depend on educated, healthy, and empowered populations. Social and cultural well-being create the foundation upon which environmental stewardship and economic growth can thrive together. Without this foundation, sustainability efforts risk becoming fragmented or short-lived.

## References

Binagwaho, A., & Scott, K. W. (2015). Improving the world's health through the post-2015 development agenda: Perspectives from Rwanda. *International Journal of Health Policy and Management, 4*(4), 203–205. https://doi.org/10.15171/ijhpm.2015.46

UNESCO. (2017). *617 million children and adolescents not getting the minimum in reading and math.* United Nations Educational, Scientific and Cultural Organization. https://www.unesco.org/en/articles/617-million-children-and-adolescents-not-getting-minimum-reading-and-math

UNESCO. (2024a). *Petra world heritage site: State of conservation report.* United Nations Educational, Scientific and Cultural Organization. https://whc.unesco.org/document/218288

UNESCO. (2024b). *Browse the lists of intangible cultural heritage and the register of good safeguarding practices.* United Nations Educational, Scientific and Cultural Organization. https://ich.unesco.org/en/lists

UNESCO. (2024c). *Out-of-school rate. Global education monitoring report.* United Nations Educational, Scientific and Cultural Organization. https://www.unesco.org/gem-report/en/view/outofschool.

UNESCO. (2025). *Data for the sustainable development goals.* UNESCO Institute for Statistics. https://databrowser.uis.unesco.org

UNHCR. (2017). *Left behind: Refugee education crisis.* The UN Refugee Agency. https://www.unhcr.org/sites/default/files/legacy-pdf/59b696f44.pdf

UNICEF. (2025). *Global education monitoring data.* United Nations Children's Fund. https://www.unicef.org/education

United Nations. (2025a). *Education for sustainable development report 2025*. United Nations. https://unstats.un.org/sdgs/report/2025/goal-04/

United Nations. (2025b). *World social report 2025: Advancing social development*. United Nations. https://www.un.org/development/desa/dspd/world-social-report.html

World Bank. (2022). *70% of 10-year-olds now in learning poverty, unable to read and understand a simple text*. The World Bank Group. https://www.worldbank.org/en/news/press-release/2022/06/23/70-of-10-year-olds-now-in-learning-poverty-unable-to-read-and-understand-a-simple-text

World Bank. (2024a). *Justice for the poor: Community empowerment and legal inclusion*. World Bank. https://www.worldbank.org/en/programs/justice-for-the-poor

World Bank. (2024b). *Gender strategy 2024–2030*. The World Bank Group. https://documents.worldbank.org/en/publication/documents-reports/documentdetail/099013107142345483

World Bank. (2025). *State of social protection report 2025*. The World Bank Group. https://www.worldbank.org/en/topic/socialprotection/publication/state-of-social-protection-2025-2-billion-person-challenge

# Chapter 4
# Economic Impacts

## 4.1 The Economic Pillar of Sustainability

Economic factors play a major role in shaping how societies function and thrive. They determine how resources are used, how wealth is distributed, and how businesses and governments respond to challenges like climate change, inequality, and resource scarcity. While the economy is often thought of in terms of profit and markets, in sustainability it is about creating systems that balance efficiency, equity, and productivity in ways that benefit both current and future generations (World Bank, 2020).

The economic pillar of sustainability emphasizes that long-term prosperity depends on using resources wisely and efficiently, building fair and inclusive opportunities, and ensuring that growth supports resilience across society. This does not mean rejecting growth or innovation. Instead, it means designing economies so that prosperity is shared more widely and natural systems can support that prosperity for decades to come (United Nations, 2023).

Economic sustainability is about building financial systems that last. This requires careful management of resources and investment in ways that avoid waste and promote long-term stability. The World Bank (2020) describes sustainable economies as those that "balance today's needs with tomorrow's opportunities," ensuring that future generations inherit both wealth and the capacity to generate it.

At its core, economic sustainability requires three elements working together:

- **Efficiency**: Making the most of scarce resources and reducing unnecessary waste.
- **Equity**: Ensuring fair opportunities for all, so growth benefits more than just a small group.
- **Productivity**: Increasing the value created from resources, labor, and innovation.

When these three principles operate in harmony, economies can grow in ways that are stable, inclusive, and environmentally responsible. Sustainable economic



N. Hugo, *Towards Sustainability*, https://doi.org/10.1007/978-3-032-29339-8_4

systems encourage innovation and diversification, helping societies adapt to change rather than collapse under it. They also promote responsible consumption and investment, linking financial decisions with long-term social and environmental goals. In this way, economic sustainability is not simply about profit, as it is about creating enduring prosperity that supports people and the planet alike.

## 4.2 Efficiency and Resource Scarcity

One of the most pressing challenges in economic sustainability is scarcity. Resources such as water, energy, and minerals are limited, and how societies use them determines whether they will be available in the future. Efficiency, which refers to getting more output from fewer inputs, helps address scarcity while saving money.

For example, global food waste reached an estimated 1.05 billion tons in the year 2022 (UNEP, 2024). This comes out to more than 1 billion meals wasted per day. Tackling inefficiency in supply chains, storage, and consumption can simultaneously reduce hunger and save enormous costs for businesses and governments.

South Korea offers a compelling example of how policy, technology, and citizen participation can dramatically reduce food waste. By enacting a "pay-as-you-throw" waste fee system and banning direct landfilling of food scraps, the country created strong economic incentives for households to sort and reduce waste. As a result, food-waste recycling rates rose from just a few percent in the 1990s to around 95% today (The Washington Post, 2024; World Economic Forum, 2019). The separated food waste is then processed into compost, animal feed, or bioenergy, reducing landfill burden and returning nutrients to agriculture.

Parma, Italy has become a leading model in waste management by implementing a comprehensive door-to-door collection system, incentive-based waste pricing, and enforcement measures including fines for improper disposal. Over roughly a decade, the city's separated collection rate for recyclable and organic waste soared from less than 50% to over 80%, dramatically reducing residual garbage and increasing recycling and composting (Le Monde, 2024). The program includes frequent collection, resident participation incentives, and infrastructure for composting, illustrating how municipal-level interventions can substantially curb food and household waste while promoting circular economy practices.

## 4.3 Equity and Fair Distribution

Equity in the economy means ensuring that prosperity is not concentrated in the hands of a few, but is distributed in ways that enhance opportunities, well-being, and quality of life for all members of society. Although global wealth has increased over recent decades, inequality remains severe. Recent analyses show that global wealth is distributed very unevenly, with the richest segments of the population controlling a disproportionately large share of total wealth. It is estimated the richest 1% of adults worldwide owned about 47.5% of global wealth in 2023, illustrating the extent of global wealth concentration (Credit Suisse, 2023).

Such disparities limit social mobility, restrict access to essential services, and reduce the overall resilience of societies. Economic sustainability seeks to address this imbalance not by diminishing wealth at the top, but by creating mechanisms that expand opportunities for those at the bottom. Imbalances cannot be corrected sustainability by simply redistributing wealth through excessive taxation or giving resources away for free, but by creating mechanisms that expand opportunities for those at the bottom while maintaining incentives for innovation and investment. These investments should be in inclusive education, support for small businesses, access to finance, and targeted skill development to allow a broader portion of the population to participate in and benefit from economic growth (UNDP, 2022). Financial inclusion, in particular, plays a critical role by providing low-income households with the tools to save, invest, and build economic security.

Kenya's M-Pesa mobile banking system illustrates this impact, serving more than 50 million people across Africa and offering affordable, secure financial services. It was found that M-Pesa lifted 2% of Kenyan households out of poverty, with especially strong benefits for women and rural communities (Jack & Suri, 2011; World Bank, 2020). By enabling mobile savings, instant transfers, and access to microloans, M-Pesa allowed households to manage risk, invest in small businesses, pay school fees, and respond more effectively to emergencies, creating lasting economic stability at the local level.

These initiatives demonstrate how expanding economic opportunities and financial inclusion can strengthen communities while supporting broader development goals. By empowering individuals to manage resources, invest in businesses, and respond to challenges, such programs contribute to more resilient and equitable societies. Globally, these efforts help reduce poverty, promote stability, and create conditions where sustainable growth can be more widely shared.

## 4.4 Productivity and Long-Term Growth

Productivity refers to the capacity to generate greater value using the same or fewer resources. It is not simply a matter of working harder, but rather of working smarter. This can be done be leveraging innovation, technological advancement, and more effective systems to maximize efficiency. In economic terms, productivity growth is a key driver of long-term development, as it enables higher incomes, improved living standards, and more sustainable patterns of resource use. The Organisation for Economic Co-operation and Development (OECD, 2020) estimates that the adoption of climate-smart policies could raise global GDP by approximately 0.23% by 2040, while simultaneously reducing poverty levels and mitigating environmental degradation.

These policies include investments in renewable energy, green infrastructure, and sustainable agriculture, which contribute to economic resilience and inclusive growth. In the agricultural sector, enhancing productivity through sustainable and innovative practices is especially critical, given its dual role in ensuring food security and supporting livelihoods for billions of people. It is estimated that global food demand will increase by about 50% by 2050 due to population growth, rising incomes, and changing dietary preferences (FAO, 2021, 2023). Meeting this demand without further depleting natural resources will require a transformation in how food is produced, processed, and distributed. This includes the adoption of precision agriculture, improved irrigation systems, better soil management, and the integration of digital technologies to optimize inputs and outputs.

Furthermore, boosting productivity in agriculture is closely tied to improving access to technology, education, and financial resources, particularly for smallholder farmers who form the backbone of food production in many developing countries. Empowering these farmers through capacity building, knowledge-sharing networks, and affordable technology can significantly enhance yields while reducing environmental impacts. For instance sustainable intensification, which is producing more output per unit of land with minimal ecological footprint, can help balance economic growth with long-term environmental health.

In the broader context of economic growth, sustained improvements in productivity also drive innovation across sectors, fostering competitiveness and enabling economies to adapt to global challenges such as climate change, resource scarcity, and market volatility. Long-term growth, therefore, depends not only on expanding output but also on improving the efficiency, inclusiveness, and sustainability of production systems. By investing in research, green technologies, and human capital, societies can ensure that productivity gains contribute to enduring prosperity and resilience for future generations.

## 4.5 The Role of Governments in Sustainable Economies

Governments play a central and indispensable role in shaping the frameworks that determine the direction and pace of sustainable economic development. Through policy design, regulation, and strategic investment, they influence how resources are allocated, how markets operate, and how societies balance growth with environmental and social responsibility. Effective governance sets the foundation for economic sustainability by encouraging innovation, ensuring equity, and maintaining the stability necessary for long-term prosperity. Instruments such as tax incentives, targeted subsidies, and public–private partnerships can steer industries toward cleaner technologies, circular economy models, and low-carbon transitions. By embedding sustainability criteria into fiscal policy and national development strategies, governments can accelerate the shift toward more inclusive and efficient economic systems.

Financial innovation is a critical part of this transformation. This includes the rapid expansion of the green bond market, which are financial instruments specifically designed to fund environmentally sustainable projects. By 2022, this market had surpassed $2 trillion globally, reflecting a growing desire among investors for sustainability-linked opportunities and signaling the financial sector's alignment with climate goals (Climate Bonds, 2023; World Bank, 2022). Governments can play a catalytic role by issuing sovereign green bonds, offering guarantees to reduce investment risks, and establishing clear standards for transparency and impact measurement. Such measures not only attract private capital into green projects but also demonstrate national commitment to international agreements like the Paris Accord and the UN Sustainable Development Goals (SDGs).

Beyond financial tools, governments are responsible for creating enabling environments that foster sustainability-oriented innovation. This involves investing in education and research, modernizing infrastructure, and promoting industries that contribute to both economic growth and environmental preservation. Policies that support renewable energy, energy efficiency, sustainable agriculture, and waste management can stimulate job creation while reducing greenhouse gas emissions and resource depletion. Moreover, integrating sustainability principles into trade policies and public procurement helps create markets for eco-friendly goods and services, reinforcing the broader transition toward a green economy.

However, governments also face the complex challenge of balancing short-term political and economic pressures with long-term sustainability objectives. Short electoral cycles and immediate budget constraints can make it difficult to prioritize policies that deliver benefits over decades rather than years. Building resilience into public budgets, infrastructure, and essential services is vital to ensure that economies can withstand and recover from shocks such as pandemics, geopolitical conflicts, or financial crises (OECD, 2020). Resilient governance requires forward-looking planning, transparent institutions, and adaptive policy frameworks that can evolve with emerging risks.

Ultimately the role of governments extends beyond regulation, as they act as facilitators, investors, and stewards of collective well-being. By aligning economic incentives with environmental limits and social equity, governments can guide the global economy toward a model that is not only prosperous but also sustainable for future generations.

## 4.6 Businesses and Corporate Responsibility

Businesses are increasingly recognizing that sustainable practices are good for both profits and long-term survival. Recent surveys show businesses are rapidly adopting sustainability practices with 96% of the world's largest 250 companies now include sustainability information in their annual reports (KPMG, 2022) and 30%-56% of CEOs and business leaders reporting that their personal compensation is linked to sustainability metrics (PwC, 2025). Companies that actively reduce waste, invest in renewable and clean technologies, or implement fair labor and ethical sourcing practices often experience a range of tangible benefits. These include cost savings through improved efficiency, enhanced innovation, risk reduction, and stronger brand loyalty from increasingly conscious consumers who prefer to support responsible businesses.

In today's marketplace, sustainability is becoming a powerful differentiator that influences purchasing decisions, investor interest, and employee engagement. Some companies that prioritize environmental and social governance performance are often rewarded with better reputations. and improved access to capital, as investors recognize that sustainable operations tend to signal long-term stability. For instance, multinational companies such as Unilever and Patagonia have successfully built their entire business models around sustainability principles. They have demonstrated that environmentally and socially responsible practices can drive both growth and consumer trust. Unilever's Sustainable Living Plan has helped reduce environmental impact across its supply chain while fostering innovation in product design, and Patagonia's commitment to environmental activism and ethical manufacturing has earned it a loyal global customer base (Patagonia, 2022; Unilever, 2020).

At the same time, smaller enterprises and start-ups are emerging as key players in sustainable innovation. Many are adopting circular economy principles, developing closed-loop systems where products, components, and materials are continually reused, repaired, and recycled to minimize waste and resource extraction (United Nations, 2023). This shift from a traditional "take-make-dispose" model to one emphasizing regeneration and resource efficiency represents a transformative step toward more resilient business ecosystems. Together, these trends indicate that sustainability is no longer an optional add-on but a strategic necessity one that enables businesses to thrive economically while contributing to broader societal and environmental well-being. We will dive deeper into these approaches and how the life cycle of a product is being more closely analyzed later in this book.

## 4.7 International Trade and Global Equity

Global trade plays a critical role in promoting economic sustainability, technological advancement, and cross-border collaboration. However, despite its potential to raise living standards and foster global prosperity, trade systems often mirror and perpetuate deep-rooted inequalities between developed and developing nations (Pavcnik, 2017; UNCTAD, 2018). Historically, industrialized countries achieved rapid economic growth through resource-intensive industries that heavily exploited natural and human resources. In contrast, developing nations today face stricter environmental and trade regulations that can constrain their economic growth, even as they attempt to follow more sustainable development pathways (World Bank, 2022). This imbalance highlights the need for a more equitable and inclusive global trade system that allows emerging economies to expand without compromising environmental integrity.

Due to non-tariff trade regulations (like sanitary, technical, or safety rules), many developing countries lose a significant share of export opportunities. For example, exporters from least-developed and lower-income countries lose an estimated $23 billion per year, equivalent to around 10% of their exports to markets from the top developed countries (UNCTAD, 2016, 2017). Such restrictions not only hinder these countries from accessing lucrative global markets, but also reduce their capacity to invest in cleaner production methods and infrastructure improvements. Many developing countries depend heavily on agricultural exports for employment and income generation, meaning that unfair trade practices can exacerbate social inequality and economic dependency.

Creating fairer trade agreements that integrate both economic opportunity and environmental responsibility is essential for ensuring long-term global stability. Policies that support technology transfer, capacity building, and access to green finance can empower developing nations to adopt sustainable industrial practices without sacrificing competitiveness. Moreover, reforming international trade frameworks to reward sustainable production, such as providing incentives for low-carbon exports or reducing tariffs on environmentally friendly products, can drive positive change across the global economy.

Ultimately, achieving global equity in trade requires cooperation among governments, international organizations, and private-sector stakeholders. A fairer and more inclusive trading system would not only help bridge the development gap but also contribute to shared environmental goals, supporting a just transition toward a greener, more resilient global economy.

## 4.8 Poverty Reduction and Inclusive Economies

Sustainability cannot succeed without directly addressing poverty and inequality, which remain among the most pressing global challenges. Poverty reduction lies at the heart of the United Nations' Sustainable Development Goals (SDGs), as persistent poverty undermines social stability, economic growth, and environmental protection. More than 700 million people still live in extreme poverty, surviving on less than $2.15 a day (United Nations, 2023). This figure underscores the urgent need for systemic change to create inclusive economic systems that generate opportunities for all, not just the privileged few. Economic sustainability is therefore not merely about maintaining growth, as it is about building resilient systems that help individuals and communities escape cycles of deprivation while reducing vulnerability to external shocks such as droughts, pandemics, climate-related disasters, or inflationary pressures.

Effective poverty alleviation strategies integrate social protection, education, and job creation to promote both immediate relief and long-term empowerment. Programs that combine these elements tend to generate stronger and more sustainable outcomes. For example, Ethiopia's Productive Safety Net Program (PSNP) has provided millions of households with food or cash transfers while engaging participants in local community projects such as soil and water conservation, reforestation, and infrastructure development (World Bank, 2017). This dual-purpose approach not only reduces short-term poverty and hunger but also strengthens environmental resilience and local capacity to cope with future crises. Similar integrated models are being adopted in other countries, where social safety nets are linked to green jobs and sustainable livelihoods, ensuring that economic progress does not come at the cost of ecological degradation.

Moreover, fostering inclusive economies requires empowering marginalized groups, such as women and rural populations, through access to education, finance, and technology. Microfinance initiatives, cooperative enterprises, and digital inclusion programs have shown promise in bridging inequality gaps, enabling individuals to participate in formal markets and benefit from global economic integration. In this way, poverty reduction becomes not just a social imperative but a cornerstone of a sustainable, inclusive global economy that leaves no one behind.

## 4.9 Measuring Success Beyond GDP

Traditional indicators like Gross Domestic Product (GDP) have long been used to assess a nation's economic performance, but they capture only a narrow aspect of development. GDP measures the total value of goods and services produced within a country, yet it fails to account for how wealth is distributed, how sustainably resources are managed, or whether people's quality of life is genuinely improving. As economies evolve and environmental limits become more apparent, policymakers and researchers increasingly recognize that growth alone is not a sufficient measure of success. A country might experience rapid GDP growth while simultaneously facing rising inequality, environmental degradation, or declining well-being. Looking at indicators beyond GDP allows for a more holistic approach to analyze long-term sustainability.

To address these limitations, the United Nations Development Programme (UNDP) introduced broader frameworks such as the Human Development Index (HDI) and the Inclusive Wealth Index (IWI) to provide a more comprehensive picture of progress (UNDP, 2022). The HDI evaluates countries based on three key dimensions: health (life expectancy), education (years of schooling), and standard of living (income per capita). Meanwhile, the IWI incorporates a broader range of assets, including manufactured, human, and natural capital, to assess whether nations are maintaining or depleting their wealth over time. Together, these indices reflect a paradigm shift toward valuing human and environmental well-being alongside economic output.

By adopting multidimensional indicators like HDI and IWI, governments can make more informed decisions that balance economic efficiency with social justice and ecological integrity. These measures encourage policymakers to prioritize investments in education, healthcare, renewable energy, and resource conservation. Redefining success beyond GDP helps ensure that progress is measured not only by how much an economy grows, but by how sustainably and equitably that growth benefits people and the planet.

## 4.10 Wrap-Up

The economic pillar of sustainability represents the foundation upon which lasting prosperity and social well-being are built. It moves beyond short-term profit and growth metrics to emphasize fairness, resilience, and the responsible use of resources. A sustainable economy is one that generates wealth without undermining the environmental systems or social structures that make future prosperity possible. By integrating principles of efficiency, equity, and productivity, societies can create a balanced framework. This allows for a focus on growing economies in a way that reduces the negative impacts to people and the environment.

Economic sustainability is deeply interconnected with innovation, inclusion, and stewardship. Efficiency ensures that limited resources are used wisely, reducing waste and maximizing value across industries and communities. Equity ensures that prosperity is shared fairly, closing the gaps between rich and poor and empowering marginalized populations through access to education, finance, and technology. Productivity drives innovation, allowing economies to do more with less, adopt cleaner technologies, and transition toward renewable energy and sustainable industries. When these three principles work together, they create economies that are both dynamic and durable, capable of adapting to global challenges such as climate change, demographic shifts, and technological disruption.

The examples explored throughout this chapter illustrate how sustainable economic strategies can succeed in practice. South Korea's food waste program demonstrates that efficiency-driven policy can save millions of dollars while reducing environmental impacts. Kenya's M-Pesa mobile banking system proves that financial inclusion not only lifts families out of poverty but also fuels entrepreneurship and local resilience. Similarly, Ethiopia's Productive Safety Net Program (PSNP) shows how linking social protection with environmental restoration can combat poverty and build long-term stability. In the business world, companies like Unilever and Patagonia show that aligning profitability with purpose strengthens both brand loyalty and market competitiveness. These real-world initiatives affirm that economic sustainability is not a theoretical ideal, but an achievable model already reshaping industries and livelihoods worldwide.

Governments, businesses, and citizens each play a crucial role in advancing this transformation. Governments can foster sustainability through policy frameworks, green investment, and inclusive trade agreements that balance economic opportunity with environmental integrity. Businesses can embed sustainability into their core operations, innovating through circular economy models, renewable energy adoption, and ethical supply chains. Meanwhile, individuals contribute by making informed consumption choices, supporting responsible enterprises, and advocating for equitable policies. Together, these efforts create a reinforcing cycle where sustainable decisions lead to long-term gains for economies and societies alike.

Ultimately, economic sustainability is not about limiting growth, but about redefining it. Growth must evolve from being purely quantitative to being qualitative, with focuses on well-being, resilience, and inclusivity. It is about shaping markets so that they value natural capital, reward innovation, and provide opportunities for all segments of society. Sustainable economic systems acknowledge that prosperity cannot come at the expense of future generations or the ecosystems that sustain life.

As global challenges intensify from climate change to inequality and resource scarcity, the economic pillar of sustainability serves as a guiding framework for building a fairer, more stable, and more prosperous world. By aligning economic growth with environmental protection and social equity, societies can ensure that development today does not compromise the potential of tomorrow. In this way, the economic dimension of sustainability becomes not only a strategy for survival but a blueprint for enduring, shared prosperity.

## References

Climate Bonds Initiative. (2023). *Global state of the market report 2023: Sustainable debt.* https://www.climatebonds.net/files/documents/publications/Global-State-of-the-Market-Report-2023.pdf

Credit Suisse. (2023). *Global wealth report 2023.* UBS/Credit Suisse. https://www.ubs.com/global/en/wealthmanagement/family-office-uhnw/reports/global-wealth-report-2023/_jcr_content/mainpar/toplevelgrid/col1/innergrid_1661914820/xcol1/linklistnewlook/link.1305067577.file/PS9jb250ZW50L2RhbS9hc3NldHMvd20vZ2xvYmFsL2ltZy9nbG9iYWwtZmFtaWx5LW9mZmljZS9kb2NzL2d3ci0yMDIzLWVuLTIucGRm/gwr-2023-en-2.pdf

FAO. (2021). *The future of food and agriculture: Trends and challenges.* Food and Agriculture Organization. https://www.fao.org/3/i6583e/i6583e.pdf

FAO. (2023). *The state of food and agriculture 2023: Towards agrifood systems that support decent livelihoods and the environment.* Food and Agriculture Organization. https://www.fao.org/publications/fao-flagship-publications/the-state-of-food-and-agriculture/2023/

Jack, W., & Suri, T. (2011). *Mobile money: The economics of M-PESA.* National Bureau of Economic Research. doi:https://doi.org/10.3386/w16721.

KPMG. (2022). *The move to mandatory reporting: Survey of sustainability reporting 2022* (key findings). https://kpmg.com/xx/en/our-insights/esg/the-move-to-mandatory-reporting.html

Le Monde. (2024, September 4). *Parma, the city that declared war on garbage.* https://www.lemonde.fr/en/environment/article/2024/09/04/parma-the-city-that-declared-war-on-garbage_6724691_114.html

OECD. (2020). *Productivity growth and environmental sustainability.* Organisation for Economic Co-operation and Development. https://www.oecd.org/environment/productivity-growth-and-environmental-sustainability.htm

Patagonia. (2022). *How Patagonia's repair program drives loyalty and customer advocacy.* Retail Dive. https://www.retaildive.com/news/patagonias-repair-program-loyalty-customer-advocacy/740865/

Pavcnik, N. (2017). *The impact of trade on inequality in developing countries.* NBER Working Paper No. 23878. National Bureau of Economic Research. doi:https://doi.org/10.3386/w23878.

PwC. (2025). *28th annual global CEO survey 2025*: Reinvention on the edge of tomorrow. https://www.pwc.com/gx/en/ceo-survey/2025/28th-ceo-survey.pdf

The Washington Post. (2024, August 9). How South Korea recycles 98% of its food waste. https://www.washingtonpost.com/world/2024/08/09/south-korea-food-waste-composting/

UNCTAD. (2016, July 19). *Developing countries lose 10 percent of exports on non-tariff measures*. United Nations Trade & Development. https://unctad.org/press-material/developing-countries-lose-10-cent-exports-non-tariff-measures-unctad

UNCTAD. (2017). *At MC11, new report reveals that trade regulations are more costly for developing countries than tariffs*. United Nations Trade & Development. https://unctad.org/news/mc11-new-report-reveals-trade-regulations-are-more-costly-developing-countries-tariffs

UNCTAD. (2018, September 26). *Unbalanced trade adding to anxiety and inequality, United Nations warns*. United Nations Trade & Development. https://unctad.org/press-material/unbalanced-trade-adding-anxiety-and-inequality-united-nations-warns

UNDP. (2022). *Human development report 2022*. United Nations development programme. https://hdr.undp.org

UNEP. (2024). *Food waste index report 2024*. United Nations environment programme. https://www.unep.org/resources/publication/unep-food-waste-index-report-2024

Unilever. (2020). *Making sustainable living commonplace for 8 billion people*. https://www.unileverusa.com/news/2020/making-sustainable-living-commonplace-for-8-billion-people/

United Nations. (2023). *The sustainable development goals report 2023*. United Nations. https://unstats.un.org/sdgs/report/2023/

World Bank. (2017). *World Bank to help Ethiopia build a national safety net system as a more effective response to droughts*. https://www.worldbank.org/en/news/press-release/2017/09/14/world-bank-to-help-ethiopia-build-a-national-safety-net-system-as-a-more-effective-response-to-droughts

World Bank. (2020). *World development report 2020: Trading for development in the age of global value chains*. World Bank. https://www.worldbank.org/en/publication/wdr2020

World Bank. (2022). *Poverty and shared prosperity 2022: Correcting course*. World Bank. https://www.worldbank.org/en/publication/poverty-and-shared-prosperity

World Economic Forum. (2019, April 12). South Korea once recycled 2% of its food waste. Now it recycles 95%. *WEF Agenda*. https://www.weforum.org/agenda/2019/04/south-korea-recycling-food-waste/

# Chapter 5
# Importance of Sustainable Practices in All Industries

## 5.1 What Industries Should Incorporate Sustainable Practices?

When most people hear the word "sustainability," their minds often go straight to industries like energy, agriculture, transportation, and construction. These sectors are some of the largest contributors to pollution, greenhouse gas emissions, and resource use, so it makes sense that they receive so much attention. For example, the energy sector alone accounts for about three-quarters (73%) of global greenhouse gas emissions (UNEP, 2022). In addition, the agriculture industry uses 70% of global freshwater withdrawals and contributes significantly to deforestation (FAO, 2021).

While these industries play a central role in environmental sustainability, it is important to recognize that sustainability extends beyond them. Areas like fashion, politics, tourism, education, and healthcare also shape the way societies consume resources, distribute opportunities, and preserve cultural traditions. Each industry connects to sustainability through environmental, social, cultural, and economic linkages. Understanding these connections allows us to see sustainability not as a narrow concept but as something that touches every part of human life.

The energy sector is often the centerpiece of sustainability discussions because of its direct link to climate change. Fossil fuels such as coal, oil, and natural gas remain dominant, and they drive the majority of global emissions. While changes in this area are occurring, such as renewable energy sources making up nearly 30% of global electricity in 2022, but major transformations are still needed from all three pillars from various industries to reach global climate goals (IEA, 2022). Remembering that sustainability involves more than just the environment and the other sectors can make an impact on economic, socio-cultural and environmental initiatives is important for advancing sustainability practices globally. Let's explore some of these industries and the impact they have.



N. Hugo, *Towards Sustainability*, https://doi.org/10.1007/978-3-032-29339-8_5

**Agriculture** Agriculture presents both challenges and opportunities. The food systems contribute one-third of global greenhouse gas emissions (FAO, 2021), however sustainable farming practices like agroforestry, crop rotation, and precision irrigation can both reduce emissions and improve productivity. Sustainable agriculture is also key for achieving food security as the world's population is projected to reach 9.7 billion by 2050 (FAO, 2021).

**Transportation** Transportation accounts for about 20% of global $CO_2$ emissions (World Bank, 2022). Expanding sustainable public transit, adopting electric vehicles, and investing in pedestrian- and bike-friendly infrastructure are all critical strategies. Cities like Bogotá, Colombia, have successfully implemented bus rapid transit systems, reducing both congestion and emissions (World Bank, 2022).

**Construction and Architecture** The construction industry is responsible for nearly 40% of global $CO_2$ emissions when including building operations and materials (UNEP, 2020). Sustainable building design, energy-efficient retrofits, and low-carbon materials such as cross-laminated timber are now being adopted as solutions. For example, Singapore's "Green Mark" building certification system has set international standards for sustainable construction (Krausmann & Musango, 2017; UNEP, 2020).

**Fashion and Merchandising** The fashion industry is often overlooked in sustainability discussions, but it has massive social, cultural, and economic impacts. The UN Alliance for Sustainable Fashion reports that the sector contributes up to 10% of global carbon emissions and produces 20% of global wastewater (UNEP, 2019). Fast fashion, with its emphasis on cheap and disposable clothing, drives overconsumption and places pressure on supply chains.

Fashion also has strong ties to culture and identity. Sustainable fashion movements highlight traditional textiles, local craftsmanship, and circular economy models, which not only reduce environmental impacts but also preserve cultural heritage. For example, initiatives in India support handloom weaving communities, giving artisans fair wages while promoting low-impact production (UNEP, 2019).

Outdoor clothing brand Patagonia has built its business model around sustainability, encouraging customers to repair clothing instead of buying new items. Its "Worn Wear" program exemplifies how fashion companies can combine profitability with circular economy principles (Patagonia, 2022).

**Political Policy** Political science and public policy play a fundamental role in sustainability. Governments shape how resources are used and how sustainability is prioritized. Policies on trade, taxation, infrastructure, and social programs all have ripple effects across industries.

The World Bank (2020) highlights that countries with stronger governance and transparent institutions experience faster poverty reduction and more resilient growth. Policies that promote equity, such as progressive taxation or investments in healthcare and education, can reduce social inequality while supporting long-term stability.

The European Union's **Green Deal** is a prime example of political leadership in sustainability. This policy framework aims to make Europe climate-neutral by 2050 while also addressing social justice through mechanisms like the Just Transition Fund, which supports workers and regions most affected by the move away from fossil fuels (European Commission, 2020).

**Tourism** Tourism contributes to cultural exchange, economic growth, and job creation, but without sustainable management it can harm ecosystems and local communities. The UN World Tourism Organization (UNWTO) reports that tourism accounts for 10% of global GDP and 1 in 10 jobs worldwide (UNWTO, 2022). However, mass tourism often leads to overcrowding, strain on local infrastructure, and cultural disruption.

Sustainable tourism promotes respect for local traditions, fair wages for workers, and reinvestment into local communities. For example, in Costa Rica, eco-tourism initiatives have created jobs while protecting cultural and natural heritage (UNWTO, 2022). This type of tourism not only strives to be low-impact on the environment, but the aim is to educate tourists in order to encourage others to continue to be mindful of their actions when they return home.

Another example is in Bhutan where they've implemented a deliberate "high-value, low-impact" tourism policy to protect its cultural heritage and environment while ensuring tourism benefits society. Under this model the Royal Government charges a Sustainable Development Fee of $100 per person, per night for most international visitors (with different arrangements for nationals of India, Bangladesh and the Maldives). That fee is used to support national priorities such as conservation, infrastructure, and social services. By curating visitor numbers and requiring tourists to use licensed tour operators, Bhutan directs tourism revenues into community and national development goals rather than mass tourism growth (Bhutan Tourism, 2024).

**Schools and Universities** Education is essential for promoting sustainable practices across all industries. Schools and universities not only teach sustainability concepts but also model them through their own operations.

The World Bank (2018) emphasizes that education is the most powerful tool for reducing poverty, with each additional year of schooling raising a person's income by 10% on average. Beyond economics, schools also shape cultural attitudes toward sustainability. Universities across the globe are integrating sustainability into curricula, from engineering to business, ensuring that future leaders are equipped to make sustainable decisions.

**Healthcare** Healthcare systems intersect with sustainability in many ways, from equitable access to care to the efficient use of resources. The World Health Organization (WHO) notes that climate change is expected to cause 250,000 additional deaths per year between 2030 and 2050 due to malnutrition, malaria, diarrhea, and heat stress (WHO, 2018).

Sustainable healthcare means more than reducing hospital energy use. It also involves ensuring access to essential care, reducing inequities, and strengthening resilience against crises like pandemics. For example, investment in universal health coverage provides long-term economic benefits by keeping workforces healthy and productive (World Bank, 2025).

Rwanda has become a model for low-income countries by expanding universal health coverage. Today, over 90% of the population has access to healthcare, supported by community-based insurance schemes. This approach has dramatically reduced child mortality and improved overall well-being, showing how healthcare sustainability can transform societies (Republic of Rwanda, n.d.).

## 5.2 Cross-Industry Collaboration

No single industry can achieve sustainability on its own. The complex, interconnected nature of modern economies means that progress in one sector often depends on cooperation and innovation across others. Cross-industry collaboration therefore represents one of the most powerful tools for accelerating sustainable development. When sectors share knowledge, technology, and investment, they can create solutions that no single organization could achieve in isolation. These partnerships not only enhance efficiency but also foster resilience by diversifying resources and expertise across industries.

For instance, the World Bank's Climate-Smart Agriculture Program demonstrates how collaboration between farmers, governments, research institutions, and technology companies can drive sustainable transformation. Through this initiative, agricultural producers gain access to digital tools and data analytics that optimize water use, improve soil health, and increase crop yields while lowering greenhouse gas emissions. Partnerships with private technology firms have enabled precision-farming innovations, such as remote-sensing systems and AI-driven irrigation controls, that make agricultural production both more profitable and environmentally responsible (World Bank, 2020).

Similarly, cross-sector collaboration during the COVID-19 pandemic highlighted the importance of linking industries that might not traditionally work together. Tourism boards, public-health agencies, and digital technology providers cooperated to design health-safety certification systems, contact-tracing platforms, and traveler-screening measures. These efforts helped destinations reopen safely and rebuild consumer confidence, protecting both lives and livelihoods (World Bank, 2020). This model has since inspired broader partnerships between the tourism, healthcare, and technology sectors to develop long-term frameworks for health-secure travel and climate-resilient destinations.

Another strong example is the energy-transportation nexus, where collaboration between automakers, energy providers, and governments is crucial to decarbonizing mobility systems. The rapid rise of electric vehicles (EVs) relies not only on innovations in car manufacturing but also on clean energy grids, battery-recycling infrastructure, and public-sector incentives. Projects such as the EV30@30 Campaign, launched by the Clean Energy Ministerial, unite stakeholders from transportation, energy, and finance to achieve a shared goal, ensuring that at least 30% of all new vehicle sales are electric by 2030 (IEA, 2022). These coordinated actions exemplify how multiple industries can align around climate targets, advancing both environmental and economic sustainability.

In construction and architecture, partnerships with the energy and manufacturing industries are driving new standards for green building design. For example, integrated projects that connect material suppliers, engineers, and renewable-energy providers are producing net-zero buildings that generate as much energy as they consume. Cities such as Copenhagen and Singapore have implemented multi-stakeholder initiatives combining government planning, private investment, and academic research to scale up green infrastructure (UNEP, 2020). Such collaborations demonstrate that the pathway to sustainable urbanization requires shared commitment across multiple sectors.

Collaboration is also reshaping the fashion industry, traditionally criticized for its environmental footprint. Apparel companies are now partnering with waste-management, technology, and logistics firms to create circular textile systems that recover, recycle, and repurpose materials. Initiatives like the Ellen MacArthur Foundation's "Make Fashion Circular" program bring together designers, retailers, and manufacturers to close the loop on textile production. This approach reduces waste, conserves resources, and inspires innovation in product design, logistics, and marketing (Ellen MacArthur Foundation, 2023).

Cross-industry collaboration further extends to finance, which underpins nearly all sustainability transitions. Banks and investment institutions are working alongside environmental scientists and policymakers to develop green financing instruments such as sustainability-linked bonds and climate-risk disclosure standards. These financial innovations enable other sectors, such as energy, agriculture, manufacturing, and construction, to access capital for low-carbon projects. By translating sustainability goals into measurable financial incentives, the finance industry becomes a critical partner in driving real-world change (World Economic Forum, 2022).

Ultimately, cross-industry collaboration reflects the understanding that sustainability is a shared responsibility and a collective opportunity. Every sector, whether public or private, influences the planet's ecological balance and social progress. By combining expertise and aligning goals across industries, societies can achieve transformative outcomes which include reducing emissions, creating jobs, improving health, and strengthening economies simultaneously.

## 5.3 Barriers and Challenges

Despite substantial progress in promoting sustainable practices, many industries still encounter significant barriers that slow the adoption of environmentally and socially responsible approaches. One major challenge is the high upfront cost associated with transitioning to greener technologies and processes. For example, retrofitting factories for energy efficiency, installing renewable energy systems, or implementing circular production models often requires substantial capital investment. For small- and medium-sized enterprises, these costs can be prohibitive, limiting their ability to innovate even when long-term savings and environmental benefits are clear.

Another critical barrier is technical knowledge and capacity. Implementing sustainability often requires specialized skills, from energy management and waste reduction to sustainable supply chain design and digital monitoring tools. Many industries lack access to the expertise, training, and advisory services needed to execute these changes effectively. Without proper guidance and technical support, companies may struggle to identify cost-effective solutions or measure the environmental and social impact of their initiatives.

Political and regulatory resistance also plays a role. In some countries, entrenched industries or policymakers may resist stricter environmental regulations or incentives for green innovation due to concerns over short-term economic disruption, job losses, or international competitiveness. Inconsistent policies and weak enforcement mechanisms further complicate long-term planning for sustainability.

The International Labour Organization (ILO) highlights both the risks and opportunities associated with these transitions. The ILO warns that moving toward a green economy could initially result in approximately 6 million job losses globally, particularly in sectors dependent on fossil fuels and carbon-intensive processes (ILO, 2018). Mining, oil extraction, coal-fired power, and heavy manufacturing are examples of industries likely to experience workforce reductions during the shift to cleaner alternatives. These disruptions can exacerbate social inequality if affected workers lack access to retraining or alternative employment opportunities.

However, the ILO also underscores the transformative potential of sustainability-driven industries. It estimates that 24 million new jobs could be created worldwide through investment in renewable energy, energy efficiency, sustainable agriculture, eco-friendly manufacturing, and green services. Renewable energy alone, which includes solar, wind, and bioenergy, has proven to be highly labor-intensive during installation, operation, and maintenance phases, providing substantial employment opportunities. Similarly, sustainable agriculture practices, such as agroforestry, organic farming, and soil restoration, require local labor and technical oversight, supporting rural economies while promoting ecological resilience.

In practice, these figures demonstrate that the transition to sustainability is not zero-sum. While some traditional jobs may decline, coordinated policy measures, vocational training, and public-private partnerships can facilitate workforce reallocation, minimize social disruption, and maximize economic and environmental gains. Industries that proactively plan for workforce transitions through reskilling programs, incentives for green entrepreneurship, and inclusive labor policies can ensure that the shift to sustainability benefits both the planet and people.

## 5.4 Wrap-Up

Sustainability is not limited to industries like energy or agriculture, even though they may be the first that come to mind. Fashion, political policy, tourism, education, and healthcare all demonstrate strong linkages to sustainability through economic, social, and cultural dimensions. Each sector plays a vital role in shaping the future, and together they reveal that sustainability is an interconnected effort. By recognizing these connections, industries can move beyond narrow views of sustainability and embrace a holistic approach that balances growth, fairness, and cultural preservation. Ultimately, the importance of sustainable practices across industries lies in their power to create systems that are more resilient, inclusive, and prepared for the challenges of tomorrow.

When we examine all of these industries together, a clear pattern begins to emerge. Each sector influences others and progress in one area often creates ripple effects that support sustainability elsewhere. For example, when political leaders adopt long term climate strategies, industries like construction, transportation, and agriculture are encouraged to innovate and reduce their environmental impacts. Similarly, when schools and universities integrate sustainability into education, they help create a generation of workers and leaders who understand the importance of social responsibility, cultural preservation, and environmental protection. Even small changes in one industry can unlock larger shifts in another, demonstrating how deeply connected these systems are.

Another key insight is that sustainability cannot be accomplished in these industries through environmental actions alone. While reducing pollution or conserving resources is essential, these efforts must be paired with economic stability, cultural respect, and social equity. A sustainable tourism program, for example, cannot succeed if it protects forests but overlooks the needs of local workers. A sustainable healthcare system cannot be effective if it reduces waste but remains inaccessible to the most vulnerable communities. Thinking in terms of all three pillars ensures that solutions are both effective and fair, helping to prevent unintended harms that might arise from one dimensional approaches.

Sustainable development becomes much stronger when industries communicate, share ideas, and learn from one another's successes and failures. Cross industry collaboration helps identify gaps, avoid duplication of effort, and create solutions that are scalable across regions and sectors. It also helps break down the false idea that sustainability is the responsibility of only a few high profile industries. Instead, every sector, from fashion to finance to education, plays a part in shaping a sustainable future.

As we move forward, the challenge will be to continue strengthening these connections and ensuring that industries are willing to adapt to changing needs. With thoughtful policies, inclusive planning, and ongoing education, sustainability becomes not just a goal, but a shared pathway for improving quality of life for present and future generations.

## References


Bhutan Tourism. (2024). *The sustainable development fee*. Tourism Council of Bhutan. https://bhutan.travel/journal/editorial/bhutan-s-sustainable-development-fee

Ellen MacArthur Foundation. (2023). *Circular fashion: Circular economy for the fashion industry*. Ellen MacArthur Foundation. https://www.ellenmacarthurfoundation.org/topics/fashion/overview

European Commission. (2020). *The European green deal*. https://commission.europa.eu/strategy-and-policy/priorities-2019-2024/european-green-deal_en

FAO. (2021, November 6). *COP26: Agricultural expansion drives almost 90 percent of global deforestation*. Food and Agriculture Organization of the United Nations. https://www.fao.org/newsroom/detail/cop26-agricultural-expansion-drives-almost-90-percent-of-global-deforestation/

IEA. (2022). *World energy outlook 2022*. International Energy Agency. https://www.iea.org/reports/world-energy-outlook-2022

ILO. (2018). *World employment and social outlook 2018: Greening with jobs*. International Labour Organization. https://www.ilo.org/global/research/global-reports/weso/greening-with-jobs/lang%2D%2Den/index.htm

Krausmann, A., & Musango, J. K. (2017). Green buildings in Singapore; Analyzing a frontrunner's sectoral innovation system. *Sustainability, 9*(6), 919.

Patagonia. (2022). *Worn wear program*. Patagonia. https://wornwear.patagonia.com

Republic of Rwanda. (n.d.). Social transformation. https://www.gov.rw/highlights/social-transformation

UNEP. (2019). *UN alliance for sustainable fashion*. United Nations Environment Programme. https://www.unep.org/news-and-stories/press-release/un-alliance-sustainable-fashion-address-environmental-impacts

UNEP. (2020). *2020 Global status report for buildings and construction*. United Nations Environment Programme. https://globalabc.org/resources/publications/2020-global-status-report-buildings-and-construction

UNEP. (2022). *Emissions gap report 2022*. United Nations Environment Programme. https://www.unep.org/resources/emissions-gap-report-2022

UNWTO. (2022). *Tourism highlights 2022*. UN World Tourism Organization. https://www.unwto.org/tourism-data/unwto-tourism-highlights

WHO. (2018). *Climate change and health*. World Health Organization. https://www.who.int/news-room/fact-sheets/detail/climate-change-and-health

World Bank. (2018). *World development report 2018: Learning to realize education's promise*. World Bank. https://www.worldbank.org/en/publication/wdr2018

World Bank. (2020). *World development report 2020: Trading for development in the age of global value chains*. World Bank. https://www.worldbank.org/en/publication/wdr2020

World Bank. (2022). *Transport global practice report 2022*. World Bank. https://www.worldbank.org/en/topic/transport/publication/transport-global-practice-annual-report-2022

World Bank. (2025, April 17). *Investing in health creates jobs and strengthens economies*. https://www.worldbank.org/en/topic/health/brief/health-economic-growth-and-jobs

World Economic Forum. (2022, November 11). *What are sustainability linked bonds and how can they support the net-zero transition?* World Economic Forum. https://www.weforum.org/stories/2022/11/cop27-sustainabilitylinked-bonds-net-zero-transition/

# Chapter 6
# Developed vs. Developing Nations

## 6.1 Sustainability and Global Equity

One of the most complex and pressing issues in sustainability is the relationship between developed and developing nations. While all countries face the impacts of climate change and environmental degradation, their capacity to respond and adapt varies widely. **Developed nations** are countries with advanced economies, strong infrastructures, and high standards of living that have already undergone extensive industrial growth. **Developing nations** are countries that are still building their economic systems, infrastructures, and social services as they work toward higher levels of stability and prosperity.

Developed countries, which have already gone through long periods of industrialization, possess infrastructures, technologies, and financial resources that allow them to implement mitigation strategies more effectively. Industrialization in these nations involved rapid economic and technological growth, often powered by fossil fuels, large-scale extraction of natural resources, and intensive land-use changes. These activities fueled economic development but came at a high environmental cost, including significant greenhouse-gas emissions, deforestation, soil degradation, and air and water pollution. By contrast, many developing countries are still in the process of industrializing. They face the dual challenge of promoting economic development and improving living standards while adhering to increasingly stringent global environmental standards and expectations that did not constrain earlier industrialization phases.

This situation creates a profound dilemma: *how can developing nations pursue economic growth while simultaneously meeting environmental regulations that their wealthier counterparts did not face during their own industrialization?*

Scholars often describe this as the "climate justice dilemma," highlighting the tension between historical responsibility and contemporary environmental urgency (Okereke, 2010; Roberts & Parks, 2007). Developing nations argue that they should



N. Hugo, *Towards Sustainability*, https://doi.org/10.1007/978-3-032-29339-8_6

not be penalized for following a developmental trajectory similar to that of early industrializers, yet the global climate crisis demands rapid emissions reductions from all countries, regardless of historical contributions. This tension underscores the importance of integrating ethical considerations, equity, and responsibility into global climate governance.

## 6.2 Historical Context of Industrialization

Industrialization in Europe, North America, and later in Japan created the conditions for these regions to develop strong economies and advanced technologies. Coal powered factories, mechanized farming, and large-scale extraction of resources made it possible for these countries to grow quickly. Because this period lasted for more than a century, these early industrializers now account for a large share of historical carbon dioxide emissions. The United States and the European Union together are responsible for approximately 40% of cumulative fossil-fuel $CO_2$ emissions since 1850, highlighting the significant historical contribution of early industrialized nations to global climate change (Global Carbon Budget, 2024).

During this early phase of industrial growth, environmental rules were limited or did not exist at all. Many governments viewed pollution as a normal part of development, and protecting the environment was not a major priority. Over time, the economic gains from industrialization allowed these countries to improve infrastructure, expand education, strengthen healthcare systems, and invest in new technologies. These long-term investments contributed to the economic advantages that industrialized nations continue to hold today.

Another important part of this history involves the global distribution of resources. During the same period, many regions in Africa, Asia, and Latin America were under colonial rule. European powers often relied on resources extracted from these territories to support their own industrial growth. This meant that raw materials such as minerals, metals, and agricultural products were taken from colonies, while local populations had limited opportunities to develop their own industries (Carbon Brief, 2023; Foreign Affairs, 2020). Therefore, historical emissions and economic development cannot be separated entirely from the colonial systems that shaped global trade and resource use.

Today developing countries face very different circumstances. They are encouraged to use cleaner and more efficient technologies, which are often more expensive and harder to access. At the same time, they must address ongoing challenges such as poverty, housing shortages, and limited infrastructure (Sachs, 2015). This creates a difficult situation in which countries must pursue economic development while following environmental expectations that earlier industrializers did not experience. How are these countries going to be able to combat all of these obstacles at once?

To help address these differences, international discussions often focus on climate finance. This includes funding, technology transfer, and training to support sustainable development in lower income countries. The United Nations estimates that hundreds of billions of dollars in climate finance will be needed each year to help developing nations adapt to climate change and reduce emissions (United Nations, 2024). Although progress has been made, many countries argue that current financial support is not yet enough to meet their development needs. Understanding these financial discussions helps explain why global climate negotiations often focus on questions of fairness, responsibility, and shared goals for the future.

## 6.3 Globalization and Unequal Growth

**Globalization**, the process through which countries become increasingly connected by exchanging goods, services, ideas, technology, and cultural practices, has significantly shaped the relationship between developed and developing nations. It has deepened these connections through expanded trade, increased investment, and greater movement of labor across borders. International supply chains frequently rely on developing countries to provide raw materials, agricultural commodities, and low-cost manufactured goods, while advanced economies capture higher-value segments, including design, branding, research and development, and high-tech services (Gereffi, 2018). This division of labor allows global companies to benefit from lower production costs in developing countries while maintaining control over the most profitable segments of the supply chain.

While globalization has generated economic growth in many regions, it also produces uneven benefits. Developing countries frequently experience the social and environmental costs of production, including pollution, deforestation, and poor labor conditions, while receiving only a small share of overall profits. Resource-rich developing regions, including many African countries, often capture a smaller share of the economic value from global mining and commodity supply chains compared to wealthier, industrialized nations that dominate high-value segments such as processing, refining, and trade (Rodrik, 2018; UNCTAD, 2020). Similarly, in Southeast Asia, textile workers may produce clothing for major international brands, but often earn wages that are insufficient to meet basic living standards, highlighting the economic disparities embedded in global trade (ILO, 2019). Yet, no international standards are in place to ensure fair wages. With issues with developed countries, such as the US, experiencing inflation and a rise in costs, many consumers are looking for the cheapest item available, regardless of quality, environmental impact, or conditions of the factories in which they were produced.

Globalization has contributed to poverty reduction in countries such as China and India, where integration into international markets has created jobs and raised incomes. However, it has also reinforced structural inequalities between nations, making it difficult for developing countries to build sustainable industries or invest in renewable energy, education, and infrastructure (Rodrik, 2018). Access to advanced technologies, capital, and global markets tends to favor developed nations, which can leverage these advantages to maintain higher productivity and innovation levels. In contrast, developing countries face challenges in upgrading their industrial base or retaining value from their resources.

These patterns have implications for sustainability. Environmental degradation and social vulnerabilities in resource-exporting countries are often linked to global demand for products consumed in wealthier nations. For instance, deforestation in the Amazon is partly driven by global demand for beef and soy, while artisanal mining in Central Africa provides critical minerals for electronics used worldwide (World Bank, 2021). Global sustainability initiatives, therefore, must consider these structural inequities to ensure that efforts to reduce emissions, conserve ecosystems, and promote fair labor practices do not disproportionately burden countries that have contributed least to historical environmental problems. Achieving equitable sustainability requires policies that support technology transfer, fair trade, climate finance, and capacity-building in developing nations, allowing them to pursue growth without repeating the environmentally harmful patterns of earlier industrialization (Sachs, 2015; UNEP, 2022).

The global trade and investment framework often reinforces this imbalance. Developed countries benefit from preferential access to markets, advanced technologies, and institutional mechanisms that allow them to maximize value from global supply chains, while developing countries struggle to retain wealth and upgrade their industrial capacity. As a result, global sustainability efforts must account for these structural inequities to avoid exacerbating economic and environmental vulnerabilities.

## 6.4 Environmental Regulations

In recent decades, international and domestic policies have introduced increasingly strict environmental regulations aimed at mitigating climate change, reducing pollution, and promoting sustainable land use. These include legally binding commitments on greenhouse-gas emissions, international rules on deforestation and land degradation, and standards for waste management and water quality. Developed nations, with their advanced economies and technological capabilities, are generally better positioned to comply with these regulations. They possess access to renewable energy technologies, advanced infrastructure, and institutional frameworks that facilitate enforcement and innovation (OECD, 2021).

Developing countries, however, face significant hurdles in implementing similar measures. Transitioning from coal, oil, and other fossil fuels to renewable energy sources requires substantial capital investment, technical expertise, and policy coordination. Global energy-related $CO_2$ emissions rose by about 1.1% in 2023, reaching a record level of approximately 37.4 billion tons (IEA, 2023). This trend of increased $CO_2$ emissions demonstrates the mounting challenges the world faces, but developing nations are at a significant disadvantage as they lack the strong economy of developed nations to advance while limiting their emissions. Without external financial and technical support, developing nations risk falling further behind economically while facing climate impacts such as extreme weather, droughts, and rising sea levels.

## 6.5 The Paris Agreement and Global Commitments

The **Paris Agreement** of 2015 marked a pivotal milestone in global climate governance. Nearly every country in the world committed to limit global warming to well below 2 °C (3.6 °F), with an aspirational target of 1.5 °C (2.7 °F) above pre-industrial levels (UNFCCC, 2015). A central principle of the agreement is "common but differentiated responsibilities and respective capabilities," which acknowledges that while all nations share responsibility for addressing climate change, wealthier nations must take the lead due to their historical emissions and greater capacity to act (Rajamani, 2016).

As part of this commitment, developed nations pledged to mobilize $100 billion annually in climate finance to support adaptation and mitigation efforts in developing countries. However, UNFCCC reports indicate that actual disbursements have frequently fallen short, contributing to mistrust and skepticism regarding global cooperation (OECD, 2023; UNFCCC, 2021). China provides an instructive example of both opportunity and challenge in sustainable industrialization. Since the 1980s, China has lifted over 800 million people out of poverty through rapid industrial growth (World Bank, 2020). However, this economic transformation came at significant environmental cost, making China the world's largest carbon emitter. More recently, China has emerged as a global leader in renewable energy deployment, producing over 70% of the world's solar panels and investing heavily in wind, hydro, battery storage, and grid modernization (IEA, 2021; Wood Mackenzie, 2023). This demonstration of change is what the Paris Agreement strives for.

However, the Paris Agreement only provides a benchmark for countries to aspire to, but unfortunately it does not have binding penalties or sanctions for countries who do not meet their emissions targets. Each country is also responsible for deciding their own commitments, which replies on peer pressure and reputation incentives for countries to set adequate goals. Therefore, the Paris Agreement is merely a platform for countries to strive to improve, but it does not have the teeth to enforce emission regulations.

## 6.6 Balancing Growth and Responsibility

Developing nations argue that they should not be held to the same environmental standards as developed countries, given their limited contributions to historical emissions. At the same time, developed nations emphasize the urgency of global climate action. This tension often surfaces in international negotiations (Roberts & Parks, 2007).

A balanced approach to this challenge requires recognizing the legitimacy of both perspectives. Developing nations need flexibility to pursue economic growth, improve infrastructure, reduce poverty, and enhance social well-being. However, this growth must be guided by strategies that avoid the environmentally destructive practices used by early industrializers. Without this consideration, developing countries risk contributing to continued global emissions increases, which can exacerbate climate impacts like flooding, heatwaves, and food insecurity, disproportionately affecting vulnerable populations (Sachs, 2015).

One key strategy for achieving this balance is international collaboration. Technology transfer is critical, allowing developing nations to adopt renewable energy systems, energy-efficient manufacturing processes, and sustainable agricultural practices without incurring prohibitive costs. For instance, solar and wind technologies, which were once costly and accessible mainly to wealthy nations, have become more affordable through international investment and innovation sharing (IEA, 2022a). Partnerships between governments, international organizations, and private companies also facilitate knowledge exchange and skills development, building local capacity to implement and maintain sustainable technologies.

Climate finance is another essential element of equitable growth. Developed nations pledged $100 billion per year under the Paris Agreement to support mitigation and adaptation efforts in developing countries. These funds help finance renewable energy projects, strengthen disaster resilience, and support reforestation and sustainable land management (OECD, 2023; UNFCCC, 2021). While implementation has sometimes fallen short, these investments provide a foundation for closing the gap between economic development and environmental responsibility.

In addition, market mechanisms such as carbon trading and carbon pricing offer opportunities for developing nations to participate in global climate solutions while generating revenue. By assigning financial value to emission reductions, these systems incentivize cleaner production methods and create new economic opportunities, including green jobs and sustainable industry growth (World Bank, 2021).

Ultimately, balancing growth and responsibility requires long-term planning, equitable partnerships, and international cooperation. Developing nations can advance economic and social development without repeating the environmentally harmful patterns of early industrializers, while developed nations can provide support, share technologies, and finance sustainable development initiatives. This shared approach is crucial for meeting global climate goals, promoting social equity, and ensuring a sustainable future for all nations (Rajamani, 2016; Sachs, 2015).

## 6.7 Utilizing Technology for Local Needs

Sustainable growth in developing communities often depends on how technology is utilized to match local conditions. Rather than relying on large, complex, and expensive systems developed in industrialized countries, many successful approaches involve smaller-scale, cost-effective solutions that align with local environmental, economic, and social realities.

In many regions, these technologies help bridge the gap between development and sustainability. For example, solar-powered lighting and small-scale renewable energy systems provide electricity in off-grid rural areas, small-scale irrigation supports local agriculture, clean cookstoves reduce indoor air pollution, and decentralized water purification systems improve access to safe drinking water. These solutions are often easier to maintain because they rely on locally available materials and skills.

A key factor in long-term success is community involvement. When local populations participate in the implementation and maintenance of technologies, including renewable energy systems, adoption rates tend to increase and systems are more likely to remain functional over time. This approach also supports job creation, skills development, and local economic growth.

However, development efforts have not always succeeded in aligning technology with local needs. In some cases, imported systems, including renewable energy infrastructure, have failed because they were too expensive to maintain, required specialized knowledge that was not locally available, or did not fit cultural practices and daily routines. Examples include large-scale water systems that fall into disrepair due to lack of maintenance capacity, or agricultural technologies that are incompatible with local farming conditions. These mismatches can lead to wasted resources, reduced trust in development initiatives, and long-term project failure.

This approach supports equitable sustainability by allowing developing nations to improve living standards while minimizing environmental harm, without repeating the resource intensive pathways of early industrializers. By prioritizing affordability, accessibility, and local relevance, these solutions help integrate economic, social, and environmental goals.

Successful technology adaptation depends on more than utilization alone. Supportive policies, education, and financing are essential, and gaps in these areas can limit long-term impact. International partnerships are therefore critical, not just for transferring technology, but for ensuring it is effectively adapted and sustained within local contexts.

## 6.8 Economic Equity and the Future of Globalization

Global renewable energy capacity continues to grow rapidly, supported by falling technology costs and policy incentives, with annual investments in clean energy exceeding $200 billion and energy storage capacity projected to surpass 2 TWh by 2030, while deployment rates vary across regions due to differences in infrastructure, supply chains, and regulatory frameworks (World Economic Forum, 2025). Clean power generation, including renewables and nuclear energy, surpassed 40% of global electricity production in 2024 (Ember, 2025). These figures highlight the rapid progress in transitioning away from fossil fuels. This is not only important for the environment, but from an economic standpoint as well, as traditional energy sources have become more expensive. The increase in gas and coal prices is responsible for about 90% of the increased cost of electricity around the world (IEA, 2022b). However, renewable energy reveals persistent regional disparities in access, investment, and infrastructure. High income countries are often able to deploy advanced energy systems more quickly, while many low and middle income nations still rely heavily on coal, oil, or biomass for electricity, limiting their ability to reduce emissions and achieve sustainable growth (IEA, 2023).

Sustainability and global equity are closely connected. Economic equity is not only a moral imperative but also a practical necessity for achieving effective climate action. If large regions remain underfunded, under resourced, and structurally disadvantaged, global sustainability initiatives risk being uneven and incomplete. Without fair access to finance, technology, and skills, developing countries may struggle to adopt cleaner technologies or implement climate adaptation measures, leaving them vulnerable to both environmental and economic shocks (Sachs, 2015).

Globalization has intensified interdependence between nations, creating both opportunities and challenges for equitable development. On one hand, international trade, foreign investment, and knowledge sharing can accelerate technology transfer, create green jobs, and support sustainable industrialization in lower-income countries. For example, multinational renewable energy projects in Africa and Southeast Asia have provided local employment while expanding access to clean electricity, illustrating how globalization can support equitable growth (UNCTAD, 2020). On the other hand, globalization can perpetuate inequalities when wealthier nations capture the highest-value segments of supply chains, leaving poorer nations to provide raw materials or low-wage labor without adequate benefits (Rodrik, 2018).

Policy frameworks and international cooperation are critical to bridging these gaps. Climate finance, capacity-building programs, and preferential access to green technologies help ensure that developing countries can participate meaningfully in the transition to a low-carbon economy. Mechanisms such as carbon markets, green bonds, and international aid can incentivize sustainable investment while supporting economic development (OECD, 2023). Moreover, equitable policies that integrate social protection, education, and gender equality contribute to long-term resilience, ensuring that no country is left behind during the shift to sustainable globalization.

Looking ahead, the future of globalization must prioritize both economic growth and environmental responsibility. By integrating justice, equity, and sustainability into trade, finance, and investment decisions, the global community can create a transition that is inclusive, effective, and resilient. This approach recognizes that climate change and economic development are interdependent, and that global challenges can only be addressed when all nations have the tools, resources, and capacity to participate fully in sustainable progress (Sachs, 2015; World Economic Forum, 2025).

## 6.9 Wrap-Up

The divide between developed and developing nations remains one of the most significant challenges in global sustainability. Developed countries benefited from industrialization powered by fossil fuels and large-scale resource extraction, building wealth, infrastructure, and technological capacity while generating substantial environmental impacts. Developing countries, in contrast, are asked to pursue economic growth under stricter environmental regulations, often without access to the same financial resources, technology, or institutional support. This imbalance reflects the historical roots of industrialization and globalization, illustrating why sustainability is not only an environmental concern but also a question of fairness and equity.

Addressing this divide requires shared responsibility. Developed nations have both the resources and the historical responsibility to provide financial assistance, technical support, and capacity-building programs to help developing countries transition to cleaner energy, adopt sustainable production practices, and build climate-resilient infrastructure. Climate finance, renewable energy partnerships, and technology transfer programs are practical tools that can enable developing nations to grow economically while limiting environmental damage (OECD, 2023; Rajamani, 2016). At the same time, developing countries must pursue development strategies that prioritize sustainable growth, invest in education and healthcare, and strengthen governance structures, ensuring that economic progress does not come at the expense of social or environmental well-being.

The Paris Agreement offers a global framework for cooperation, emphasizing the principle of common but differentiated responsibilities. It highlights the need for equitable participation and recognizes the varied capacities of nations to respond to climate change. Yet, the agreement alone is not sufficient. Success will depend on trust, transparency, and continued international collaboration, as well as policies that integrate climate action with economic and social development goals.

Ultimately, sustainability is about more than reducing emissions or protecting natural resources. It is about creating a fair and inclusive global system where all nations can thrive, and where economic, social, and environmental goals are pursued together. By bridging historical inequalities, fostering cooperation, and promoting responsible growth, the global community can build a sustainable future that benefits both people and the planet. Achieving this balance is challenging, but it is essential for long-term resilience, equity, and shared prosperity.

## References

Carbon Brief. (2023). *Revealed: How colonial rule radically shifts historical responsibility for climate change.* https://www.carbonbrief.org/revealed-how-colonial-rule-radically-shifts-historical-responsibility-for-climate-change/

Ember. (2025). *Global electricity review 2025.* https://ember-climate.org/data/global-electricity-review/

Foreign Affairs. (2020). *The climate debt the West owes the rest.* https://www.foreignaffairs.com/articles/world/2020-04-13/climate-debt

Gereffi, G. (2018). *Global value chains and development: Redefining the contours of 21st-century capitalism.* Cambridge University Press. https://doi.org/10.1177/1024529420929980

Global Carbon Budget. (2024). *Cumulative $CO_2$ emissions.* https://ourworldindata.org/grapher/cumulative-co-emissions

IEA. (2022a). *Solar PV global supply chains.* International Energy Agency https://www.iea.org/reports/solar-pv-global-supply-chains/executive-summary

IEA. (2022b). *World energy outlook 2022.* International Energy Agency https://www.iea.org/reports/world-energy-outlook-2022

IEA. (2023). *SDG7: Data and projections: Access to electricity.* International Energy Agency https://www.iea.org/reports/sdg7-data-and-projections/access-to-electricity

ILO. (2019). *Global wage report 2018/19: What lies behind gender pay gaps.* International Labour Organization. https://www.ilo.org/global/publications/books/WCMS_650553/lang%2D%2Den/index.htm

International Energy Agency. (2021). *Renewable electricity: Renewable energy market update 2021.* IEA. https://www.iea.org/reports/renewable-energy-market-update-2021/renewable-electricity

OECD. (2021). *OECD environmental performance review.* Organisation for Economic Co-operation and Development. https://www.oecd.org/en/publications/serials/oecd-environmental-performance-reviews_g1gha1c8.html

OECD. (2023). *Climate finance provided and mobilised by developed countries in 2013–2021.* Organisation for Economic Co-operation and Development https://www.oecd.org/content/dam/oecd/en/publications/reports/2023/11/climate-finance-provided-and-mobilised-by-developed-countries-in-2013-2021_517fec8e.pdf

Okereke, C. (2010). Climate justice and the international regime. *Wiley Interdisciplinary Reviews: Climate Change, 1*(3), 462–474. https://doi.org/10.1002/wcc.52

Rajamani, L. (2016). Ambition and differentiation in the 2015 Paris agreement: Interpretative possibilities and underlying politics. *International and Comparative Law Quarterly, 65*(2), 493–514. https://doi.org/10.1017/S0020589316000130

Roberts, J. T., & Parks, B. C. (2007). *A climate of injustice: Global inequality, north–south politics, and climate policy.* MIT Press. https://mitpress.mit.edu/9780262681612/a-climate-of-injustice/

Rodrik, D. (2018). *Straight talk on trade: Ideas for a sane world economy.* Princeton University Press. https://press.princeton.edu/books/hardcover/9780691177847/straight-talk-on-trade?srsltid=AfmBOooSFGJqcipsMNEIIRcSU2xr_hG8nnRNoQQp3ocVYf8cxUFjHDBH

Sachs, J. (2015). *The age of sustainable development.* Columbia University Press. https://cup.columbia.edu/book/the-age-of-sustainable-development/9780231173155/

UNCTAD. (2020). *The state of commodity dependence 2020.* United Nations Conference on Trade and Development. https://unctad.org/webflyer/state-commodity-dependence-2020

UNEP. (2022). *Emissions gap report 2022.* United Nations Environment Programme. https://www.unep.org/resources/emissions-gap-report-2022

UNFCCC. (2015). *The Paris Agreement.* United Nations Framework Convention on Climate Change https://unfccc.int/sites/default/files/english_paris_agreement.pdf

UNFCCC. (2021). *Report on the standing committee on finance 2021.* United Nations Framework Convention on Climate Change. https://unfccc.int/documents/271924

United Nations. (2024). *Key findings on climate finance.* https://www.un.org/en/climatechange/science/key-findings

Wood Mackenzie. (2023, November 15). *China's dominance in global solar supply chains continues to grow.* https://www.woodmac.com/press-releases/china-dominance-on-global-solar-supply-chain/

World Bank. (2020). *Poverty and shared prosperity 2020: Reversals of fortune.* https://www.worldbank.org/en/publication/poverty-and-shared-prosperity-2020

World Bank. (2021). *Access to electricity: Financing needs for universal energy access (Policy document).* https://documents1.worldbank.org/curated/en/099122223123530408/pdf/P174481 1ffdf0c0f1b0f71ebda9ad18828.pdf

World Economic Forum. (2025). *Fostering effective energy transition 2025: Redefining global energy systems*. World Economic Forum. https://www.weforum.org/publications/fostering-effective-energy-transition-2025/in-full/redefining-global-energy-systems/

# Chapter 7
# Sustainable Development Goals

## 7.1 What Are the Sustainable Development Goals?

The **Sustainable Development Goals (SDGs)** represent one of the most ambitious global efforts to create a fairer, healthier, and more sustainable world. They were adopted in 2015 by the **United Nations (UN)**, which is an international organization founded in 1945 and made up of 193 Member States, promotes peace, security, development, and human rights. There SDGs are comprised of17 goals in total, ranging from eliminating poverty and hunger to combating climate change and promoting peace (United Nations, 2015).

What makes the SDGs unique is their universality. Unlike past development frameworks, they apply to all countries, with developed and developing alike taking on the challenge of sustainable development. The SDGs recognize that global challenges such as poverty, inequality, climate change, and health are interconnected and cannot be solved in isolation (Sachs et al., 2019).

**Historical Background** The SDGs built upon an earlier framework known as the Millennium Development Goals (MDGs), which guided global development from 2000 to 2015. The MDGs focused on eight areas, including reducing extreme poverty, improving maternal health, and increasing access to education. While progress was significant, with global extreme poverty falling from 36% in 1990 to 10% in 2015 (World Bank, 2020), the MDGs faced criticism for being too narrow, focusing mainly on developing nations, and lacking strong environmental targets.

The SDGs expanded the scope further on social, environmental, and economic dimensions of development. They also emphasized universality, recognizing that issues like sustainable consumption, inequality, and climate change affect all countries, not just those in the developing world (United Nations, 2015).



N. Hugo, *Towards Sustainability*, https://doi.org/10.1007/978-3-032-29339-8_7

## 7.2 The 17 Sustainable Development Goals

The 17 Sustainable Development Goals address the world's most pressing challenges. Each goal has targets (169 in total) and indicators (over 230) to measure progress. The goals are designed to balance the three pillars of sustainability:

- **Economic growth** (decent work, innovation, reduced inequality)
- **Social inclusion** (education, health, gender equality, peace)
- **Environmental sustainability** (climate action, clean water, biodiversity)

The SDGs cover a wide range of issues and have linkages, meaning progress in one area often supports another. For example, improving access to education (Goal 4) contributes to reducing poverty (Goal 1) and promoting gender equality (Goal 5).

The 17 Goals are:

1. No Poverty
2. Zero Hunger
3. Good Health and Well-being
4. Quality Education
5. Gender Equality
6. Clean Water and Sanitation
7. Affordable and Clean Energy
8. Decent Work and Economic Growth
9. Industry, Innovation, and Infrastructure
10. Reduced Inequalities
11. Sustainable Cities and Communities
12. Responsible Consumption and Production
13. Climate Action
14. Life Below Water
15. Life on Land
16. Peace, Justice, and Strong Institutions
17. Partnerships for the Goals

The goals can be grouped based on the three pillars of sustainability. Below is an overview of the categories they fall into.

**Social Goals**

- *No Poverty & Zero Hunger (Goals 1 & 2):* Over 700 million people still live in extreme poverty, and nearly 9% of the global population faced hunger in 2022 (FAO, 2023). Progress requires tackling inequality, providing social safety nets, and building resilient food systems.
- *Health & Education (Goals 3 & 4):* Global child mortality has fallen by 60% since 1990, but access to quality healthcare and education remains unequal. COVID-19 disrupted schooling for 1.6 billion students worldwide (UNESCO, 2020).
- *Peace, Justice & Partnerships (Goals 16 & 17):* Achieving the SDGs requires strong institutions and international cooperation (UNDESA, 2023). Policies

provide benchmarks for society and businesses to follow, as well as ensuring compliance to mandates.

**Economic Goals**

- *Decent Work & Growth (Goal 8):* The global unemployment rate is about 5%, but underemployment and informal work remain high, especially in developing countries.
- *Industry, Innovation, Infrastructure (Goal 9):* Investments in green infrastructure could create 65 million new jobs by 2030 (ILO, 2019; OECD, 2019).

**Environmental Goals**

- *Climate Action (Goal 13):* The planet has already warmed by about 1.1 °C (2 °F) above pre-industrial levels, increasing extreme weather events (IPCC, 2021).
- *Life on Land & Below Water (Goals 14 & 15):* One million species are at risk of extinction (IPBES, 2019).

## 7.3 Highlights of SDG Progress

**Nigeria: Poverty Reduction (SDG 1)**

Nigeria faces significant challenges under SDG 1. Despite being one of Africa's most resource-rich countries, more than 90 million Nigerians live in extreme poverty, reflecting deep structural inequalities that limit broad-based development (World Bank, 2022). The National Social Safety Nets Project (NSSNP) has expanded cash transfers and support programs to vulnerable households, helping reduce immediate hardship. However, issues such as unemployment, regional instability, and gaps in infrastructure continue to constrain long-term progress. Nigeria demonstrates why poverty reduction requires not only economic growth but also strong governance and inclusive social policies that ensure resources reach those who need them most.

**Rwanda: Good Health and Well-being (SDG 3)**

Rwanda has achieved notable improvements aligned with SDG 3. Life expectancy rose from around 49 years in 2000 to nearly 70 years in 2022 due to investments in community-based health insurance, national vaccination campaigns, and stronger primary healthcare services (WHO, 2022). Expanded access to clinics and trained health workers has also reduced maternal and child mortality. Rwanda's progress shows how coordinated national policies, community participation, and international support can improve public health even in low-income settings. Although challenges remain in financing and specialist care, the country provides a strong example of how targeted interventions can produce measurable gains.

**Bangladesh: Quality Education (SDG 4)**

Bangladesh has made important gains under SDG 4, as primary school enrollment exceeds 97%, and the country has achieved gender parity in both primary and secondary education (UNESCO, 2020). Government policies such as stipends for girls, community-based schools, and free primary education have expanded access in rural and underserved regions. International partnerships have also supported curriculum reforms and teacher training. While learning quality and dropout rates remain areas for improvement, Bangladesh's progress shows how targeted strategies can broaden educational access even in resource-limited contexts.

**India: Affordable and Clean Energy (SDG 7)**
India has made major advances towards SDG 7, particularly in renewable energy development. Solar capacity increased from just 2.6 gigawatts (GW) in 2014 to more than 70 GW by 2023, making India one of the fastest-growing solar markets worldwide (IEA, 2023). The government has supported this expansion through national missions, subsidies, and large-scale solar parks. However, coal still provides most of India's electricity, reflecting the challenge of transitioning while meeting rising energy demand. India's progress illustrates both the potential for rapid renewable expansion and the complexities facing large developing economies balancing sustainability with growth.

**Vietnam: Decent Work and Economic Growth (SDG 8)**
Vietnam has made important strides under SDG 8, as the country has implemented policies to boost employment, improve labor conditions, and create higher-quality jobs through both domestic growth and foreign investment. Firms like Vu Phong Energy Group have actively contributed by expanding operations, building new offices, and generating well-paying work for local communities (Vu Phong Energy Group, 2025). These efforts help ensure that economic growth in Vietnam is more inclusive and sustainable. Despite these gains, challenges persist in ensuring labor rights, reducing underemployment, and scaling up job opportunities. Vietnam's experience shows how economic growth can be aligned with decent work when companies, government, and international partners cooperate.

**Singapore: Sustainable Cities (SDG 11)**
Singapore is a leading example for SDG 11. Through its "Smart Nation" initiative, the city-state has integrated digital technology, data, and planning to make its urban systems more efficient, inclusive, and sustainable (UNECE, 2020). The government has used smart sensors, real-time traffic management, and public transport integration to reduce emissions, improve public services, and maintain high quality of life. Singapore's long-term land-use planning, reflected in its master plans, prioritizes green spaces, safe public transport, and disaster resilience, helping the city balance growth with sustainability (Singapore Department of Statistics, 2024). This case illustrates how technology-driven urban planning can support inclusive, resilient, and environmentally responsible cities that serve the needs of all residents.

**Norway: Responsible Consumption and Production (SDG 12)**
Norway is a global leader in SDG 12, especially in recycling and circular-economy practices. The country recycles more than 95% of plastic bottles through an efficient deposit-return system that is widely supported by the public and industry (Ellen MacArthur Foundation, 2020). Strong environmental policies, clear consumer incentives, and accessible recycling infrastructure contribute to high participation

rates. Norway's approach shows that responsible consumption can be achieved when governments, businesses, and citizens work together to minimize waste and conserve resources. Its success also demonstrates how policy design can shape sustainable everyday behaviors.

**Costa Rica: Climate Action (SDG 13)**
Costa Rica has become a model for SDG 13. For several years, more than 98% of its electricity has come from renewable sources such as hydropower, wind, geothermal, and solar energy (IEA, 2022). This is in part due to the volcanic activity in the country. Costa Rica has committed to achieving carbon neutrality by 2050, supported by long-term environmental policies, availability and reliability of renewable energy sources, and strong public support for nature conservation. Although transportation emissions remain a challenge, Costa Rica's progressing towards positive change.

## 7.4 Measuring SDGs on your Own

Students and researchers can explore SDG progress using interactive, publicly available data tools from the United Nations and the World Bank. These resources allow users to access, analyze, and visualize indicators for their own research projects.

The UN SDG Global Database (https://unstats.un.org/sdgs/) compiles official SDG indicators reported by countries and UN agencies.

Key features:

- Interactive dashboards to track indicators over time.
- Disaggregated data by sex, age, income, and geography to analyze equity and inclusion.
- Downloadable datasets covering health, education, environment, and economic indicators.

Students can use this database to compare countries, explore trends over time, and create charts or graphs for reports and presentations.

The World Bank (https://data.worldbank.org) provides over 1000 indicators aligned with many SDG targets.

Highlights Include

- Direct mapping to SDG targets, such as poverty, education, energy use, water resources, and health outcomes.
- Tools for visualization and data download in multiple formats.
- API access for students interested in programming or advanced data analysis.

The World Bank allows users to examine trends, make cross-country comparisons, and integrate data with UN SDG indicators for richer research. Appendix D of this textbook gives an overview of what type of information is available and how to use it.

## 7.5 Challenges in Implementation

Despite meaningful progress in global development, significant barriers remain that threaten the achievement of the SDGs. One of the most constraining of these is financing. The United Nations reports that developing countries face an annual funding gap of around $2.5 trillion in order to meet SDG related investment needs under current conditions (United Nations, 2018). More recent analyses have pushed that figure even higher, rising to as much as $4.0 trillion -$4.2 trillion yearly when accounting for converging crises, declining investment, and slower growth (International Institute for Sustainable Development, 2024). The magnitude of this gap underscores the fact that while many countries have set ambitious goals, the financial resources needed to match those ambitions are lacking, especially in low and middle-income countries.

Another major barrier is data collection and monitoring. In many low-income nations, statistical and surveillance systems are weak or under-resourced, making it difficult to track progress reliably. Inconsistent data complicates efforts to measure whether SDG targets are being met, identify where investments are falling short, and direct resources to the most vulnerable populations. Without robust data, the risks of inefficient or mis-aligned investments increase, and the ability to hold governments and international actors accountable for progress diminishes.

Global crises compound these structural issues. The onset of the COVID-19 pandemic triggered the largest jump in global extreme poverty in decades. It is estimated that between 88 million −115 million additional people were pushed into extreme poverty in 2020 as a result of the pandemic (World Bank, 2020). Moreover, the United Nations estimated that due to the combined effects of the pandemic and other shocks, an additional 93 million people may have been living in extreme poverty in 2020 compared to pre-pandemic projections (United Nations Statistics Division, 2022). These reversals wiped out years of progress toward poverty eradication goals.

Conflicts, such as the war in Ukraine, and climate disasters further deepen the challenge. Wars disrupt food systems, displace populations, damage infrastructure, and divert national resources from development to relief and defense. Climate-related disasters, including hurricanes, floods, droughts, and wildfires, not only cause immediate human suffering but erode long-term development gains by destroying schools, health systems, homes, crops, and energy infrastructure. The cumulative effect of these overlapping crises is that progress toward the SDGs becomes slower, more expensive, and more uncertain.

In this environment of constrained resources, weak data systems, and compounded crises, achieving the SDGs by 2030 becomes more difficult. Financing shortages mean that even when good data exist and intentions are aligned, the capacity to act is inadequate. Data weakness means interventions may not be optimally targeted, slowed, or mis-directed. And global shocks create setbacks, forcing countries to reallocate resources toward recovery rather than forward-looking development. Together, these dynamics raise risks of widening inequality, stalled growth, and unmet targets, which in turn undermine confidence in the global development enterprise.

## 7.6 Partnerships, Global Cooperation and Future Outlook

Goal 17 highlights the importance of partnerships. No single country can achieve the SDGs alone. International cooperation through financing, technology transfer, and trade reform is essential. Developed nations are expected to contribute 0.7% of their Gross National Income (GNI) to development aid, but most fall short (OECD, 2022). Strengthened partnerships between governments, businesses, and civil society are necessary to mobilize resources, bridge knowledge gaps, and scale sustainable innovations.

Beyond finance, SDG 17 stresses multi-stakeholder partnering: governments must collaborate with the private sector, civil society organizations, academia and local communities to mobilize resources, bridge knowledge gaps, scale innovations, and ensure that the benefits of sustainable development are shared widely (Target 17.17). Such partnerships allow for synergies, in which private capital can fill investment gaps, businesses can provide technical solutions, research institutions can adapt technologies for local contexts, and civil society can help ensure participation, transparency and equity.

With only a few years left until 2030, the UN reports that only 15% of SDG targets are on track (UN, 2023). Among the 169 targets only 138 could be assessed, but of those nearly half (48%) are either moderately or severely off-track and over one-third (37%) show stagnation or regression compared to the 2015 baseline (United Nations, 2023).

While this is concerning, the SDGs remain a powerful global framework for action. They provide not just goals, but also indicators that enable accountability. They have already demonstrated their ability to guide businesses and governments in a direction to strive towards progress in sustainable initiatives. The indicator-based approach means that countries, institutions, and general society can see *where* progress is happening, *where* it is lagging, and have data to assist them in understanding *why*. Moving forward, success will depend on political will, equitable financing, and grassroots-level implementation that empowers local communities to take ownership of sustainability.

## 7.7 Wrap-Up

The Sustainable Development Goals represent a broad international effort to guide countries toward a more sustainable and inclusive future. They outline a shared vision that brings together economic growth, social well-being, and environmental stewardship. These three areas work in combination, and progress in one often depends on advancements in the others. Although progress across the goals varies from region to region, the SDGs have helped create a common direction for governments, institutions, businesses, and communities. This shared framework has supported policy development, encouraged collaboration, and increased attention to global challenges.

The coming years are important for turning the SDGs into measurable outcomes. Many countries continue to navigate limitations such as constrained financial resources, uneven governance capacity, and the effects of climate related events. Inequality also remains a significant barrier in many regions. These factors can slow progress, yet the examples in this chapter show that improvement is possible under the right conditions. Rwanda's progress in healthcare and Bangladesh's commitment to high quality education illustrate how thoughtful planning, public engagement, and consistent leadership can support national development goals. These examples also show that meaningful change can occur in a range of contexts, not just in countries with unusual advantages.

The SDGs also highlight that sustainability includes more than environmental concerns. It involves fair access to opportunities, stable social systems, and support for populations that face persistent disadvantages. The goals draw attention to the ways in which issues such as education, health, poverty reduction, and responsible consumption influence one another. When progress is made in one area, it often contributes to positive change elsewhere. This interconnected design encourages a more complete understanding of development and the factors that shape long term well being.

As the world moves closer to 2030, the SDGs continue to stress the value of cooperation across borders and sectors. Challenges such as climate change and economic instability extend beyond national boundaries, which means collective action is necessary. Partnerships among governments, private organizations, researchers, and community groups will remain important for supporting innovation, sharing knowledge, and implementing effective solutions. The SDGs encourage long term thinking, investments in resilience, and approaches that aim to distribute the benefits of development more broadly. While the path forward is complex, the goals provide a useful guide for planning and evaluating efforts to build a more stable and sustainable future for all.

## References

Ellen MacArthur Foundation. (2020). *Completing the picture: How the circular economy tackles climate change*. Ellen MacArthur Foundation. https://www.ellenmacarthurfoundation.org/assets/downloads/Completing-The-Picture_Circular-Economy-and-Climate-Change_V3_26_05_20.pdf

FAO. (2023). *The state of food security and nutrition in the world 2023*. Food and agriculture Organization. https://data.unicef.org/resources/sofi-2023/

IEA. (2022). *Renewables 2022: Analysis and forecast to 2027*. International Energy Agency. https://www.iea.org/reports/renewables-2022

IEA. (2023). *World energy outlook 2023*. International Energy Agency. https://www.iea.org/reports/world-energy-outlook-2023

ILO. (2019). *Skills for a greener future: A global view*. International labour Organization. https://www.ilo.org/global/publications/books/WCMS_732214/lang--en/index.htm.

International Institute for Sustainable Development. (2024). *Annual SDG financing gaps measured in trillions: FSDR 2024*. https://sdg.iisd.org/news/annual-sdg-financing-gaps-measured-in-trillions-fsdr-2024/?utm_source=chatgpt.com

IPBES. (2019). *Global assessment report on biodiversity and ecosystem services*. Intergovernmental Science-Policy Platform on Biodiversity and Ecosystem Services. https://files.ipbes.net/ipbes-web-prod-public-files/inline/files/ipbes_global_assessment_report_summary_for_policymakers.pdf

IPCC. (2021). *Climate change 2021: The physical science basis*. Intergovernmental Panel on Climate Change.

OECD. (2019). *Sustainable infrastructure for low-carbon development in Central Asia and the Caucasus: Hotspot analysis and needs assessment*. Organisation for Economic Co-operation and Development. https://www.oecd.org/en/publications/sustainable-infrastructure-for-low-carbon-development-in-central-asia-and-the-caucasus_d1aa6ae9-en.html

OECD. (2022). *Development co-operation report 2022*. Organisation for Economic Co-operation and Development. https://www.oecd.org/en/publications/serials/development-co-operation-report_g1gha553.html

Sachs, J., Schmidt-Traub, G., Kroll, C., Lafortune, G., Fuller, G., & Woelm, F. (2019). *Sustainable development report 2019*. Bertelsmann Stiftung and Sustainable Development Solutions Network (SDSN). https://sustainabledevelopment.report/2019/

Singapore Department of Statistics. (2024). *Singapore in figures 2024*. Government of Singapore. https://www.singstat.gov.sg/publications/reference/singapore-in-figures

UNDESA. (2023). *The sustainable development goals report 2023*. United Nations Department of Economic and Social Affairs. https://unstats.un.org/sdgs/report/2023/

UNECE. (2020). *Implementing smart sustainable cities: The case of Singapore*. United Nations Economic Commission for Europe. https://standards4sdgs.unece.org/sites/default/files/2020-01/SDG11_Singapore.pdf

UNESCO. (2020). *Global education monitoring report 2020*. United Nations Educational, Scientific and Cultural Organization.

United Nations. (2015). *Transforming our world: The 2030 agenda for sustainable development*. United Nations. https://sdgs.un.org/2030agenda

United Nations. (2018). *Unlocking SDG financing: Findings from early adopters*. United Nations. https://unsdg.un.org/resources/unlocking-sdg-financing-findings-early-adopters?utm_source=chatgpt.com

United Nations. (2023). *The sustainable development goals: Progress chart 2023*. United Nations. https://unstats.un.org/sdgs/report/2023/progress-chart/Progress-Chart-2023.pdf

United Nations Statistics Division. (2022). *The sustainable development goals report 2022, goal 1: Extreme poverty*. https://unstats.un.org/sdgs/report/2022/goal-01/

Vu Phong Energy Group. (2025). *SDG 8: Decent work and economic growth*. https://vuphong.com/sdg8-decent-work-and-economic-growth/

WHO. (2022). *World health statistics 2022*. World Health Organization. https://www.who.int/news/item/20-05-2022-world-health-statistics-2022

World Bank. (2020). *Poverty and shared prosperity 2020: Reversals of fortune*. World Bank. https://www.worldbank.org/en/publication/poverty-and-shared-prosperity-2020

World Bank. (2022). *Poverty and inequality platform: Nigeria country profile*. World Bank. https://pip.worldbank.org/country-profiles/IDN

# Chapter 8
# Conflicting Goals and Tradeoffs

## 8.1 What Are Tradeoffs?

Sustainability sounds simple in theory. Just balance the environment, the economy, and society so people and the planet can thrive together. However, the practice of sustainability often involves tradeoffs. A **tradeoff** occurs when making progress in one area means sacrificing progress in another. Tradeoffs focus on analyzing if the overall benefit still outweighs the cost and appear in nearly every aspect of sustainable development because global systems are interconnected. These situations reveal the complexity of global systems, where no decision happens in isolation.

For example, a government may encourage agricultural expansion to end hunger (SDG 2), but this could lead to deforestation and biodiversity loss, undermining climate goals (SDG 13 and SDG 15). Another example of a tradeoff is seen with trophy hunting, in which animals are killed for sport, but the fees collected are used to fund conservation programs, anti-poaching efforts, and community development, ultimately helping protect wildlife populations and support local livelihoods. The idea of killing animals for sport is seen as a strong negative, but as the money is used to improve conservation efforts and livelihoods of local communities, trophy hunting can help animals and community members more than it hurts it. Policymakers, businesses, and communities constantly face these choices, weighing which priorities matter most in the short and long term.

Urbanization provides another clear example. Cities generate jobs, expand educational opportunities, and support economic growth. They make it possible to deliver healthcare and public services more efficiently, because people live in closer proximity to one another. Yet rapid urban expansion can increase greenhouse gas emissions, strain water and transportation systems, and contribute to the loss of natural habitats. Policymakers must decide how to accommodate growing urban populations while also managing pollution, waste, and land use pressures. These



N. Hugo, *Towards Sustainability*, https://doi.org/10.1007/978-3-032-29339-8_8

decisions often involve prioritizing which needs require immediate attention and which concerns can be addressed through long-term planning.

Understanding tradeoffs is a key skill for anyone studying sustainability because it highlights that perfect solutions rarely exist. Instead of trying to eliminate all tradeoffs, societies aim to make informed decisions that produce the greatest overall benefit while minimizing harm. Policymakers, businesses, scientists, and communities must work together to understand the full range of consequences associated with each choice. The goal is not to avoid tradeoffs entirely but to manage them transparently and responsibly, acknowledging that every decision comes with both benefits and costs. By studying tradeoffs, we learn how to design strategies that support long-term well-being while protecting resources for future generations.

## 8.2 Balancing Impacts

A complex example of a tradeoff is between improving human health and the resulting population growth. Global life expectancy has risen substantially since 1950 and reached about 73 years by 2022 with continued gains in health and reductions in child mortality are a major success (United Nations, 2019, 2022). While this is a major success, it also means populations are growing in many regions, particularly in Africa and Asia. The UN projects the global population will reach 9.7 billion by 2050 (United Nations, 2019). More people means there is a rising demand to meet needs, as more people require more food, water, housing, energy, and healthcare. This puts an enormous strain on natural resources. As a result, difficult decisions need to be made.

Another major tradeoff occurs between feeding people and protecting the planet. Agriculture currently uses about 38% of the world's land and accounts for nearly 70% of freshwater withdrawals (FAO, 2021). At the same time, it generates about one-third of global greenhouse gas emissions (World Bank, 2022). Efforts to end hunger by intensifying farming often lead to deforestation, soil erosion, and water pollution. For instance, palm oil production in Indonesia and Malaysia created millions of jobs and lifted rural families out of poverty. However, it also destroyed vast rainforests and threatened endangered species like orangutans (Koh & Wilcove, 2008). Tradeoffs in agriculture highlight a difficult question: *do we prioritize immediate food needs or the long-term stability of ecosystems that future generations depend on*?

Many countries, especially developing ones, often face a tradeoff between economic growth and environmental sustainability. Industrialization and infrastructure projects bring jobs, boost GDP, and reduce poverty, but they frequently come at the cost of pollution, carbon emissions, and habitat destruction. China is a prime example. Its rapid economic growth lifted 800 million people out of poverty since the 1980s (World Bank, 2022). However, it also led to severe air pollution and China now produces about 30% of global $CO_2$ emissions (IEA, 2021). The tradeoff between development and environmental health is stark, but it also shows why many developing nations argue that rich countries, which polluted heavily during their own growth, should shoulder more of the responsibility for emissions reductions (Roberts & Parks, 2007).

## 8.3 Short-Term Gains Vs. Long-Term Sustainability

Many tradeoffs boil down to the tension between short-term benefits and long-term sustainability. Governments and businesses often prioritize policies that bring immediate results, such as building a new highway, opening a coal mine, or expanding industrial zones, because these actions create jobs, increase tax revenue, and boost economic growth now. However, such decisions can lock countries into unsustainable pathways. For instance, investments in fossil fuel infrastructure create "carbon lock-in," making it much harder and costlier to transition to renewable energy in the future (Unruh, 2000).

Convincing stakeholders to sacrifice immediate profits for the sake of future generations is challenging, influenced by both economic interests and human psychology. Behavioral economics shows that people often heavily favor immediate rewards over long-term benefits, a phenomenon known as "present bias" (Laibson, 1997). Overcoming this bias is essential to align development and environmental goals.

Let's go back to the idea of trophy hunting, from the beginning of this chapter. On the surface, trophy hunting seems harmful and ethically troubling because it involves the killing of iconic species like lions, elephants, and rhinoceroses. In parts of Africa, however, it has been defended as a mechanism for raising funds for wildlife conservation and providing livelihoods for rural communities. Trophy hunting contributes approximately $200 million annually to sub-Saharan economies, and some of these revenues fund anti-poaching patrols, habitat restoration, and community development projects (IUCN, 2016). Despite this, it remains highly controversial as the ethical implications and potential ecological risks continue to draw global attention (Ripple et al., 2016). Yet, alternatives to generating an equivalent amount of money to fund conservation and community projects has not been found. It is also important to note that this money tends to come from tourists from developed countries, so the injection of funds into the community helps to strengthen the overall economy without increasing taxes for locals.

Cultural heritage and modernization also highlight difficult tradeoffs. Economic expansion often requires demolishing historic buildings, displacing residents, and altering cultural landscapes. For instance, in India, development pressures threaten historic areas such as Old Delhi, where modern infrastructure projects compete with preservation efforts (UNESCO, 2020). Globally, UNESCO has warned that numerous World Heritage Sites face similar risks. Places like well-visited Venice, Italy deal with the weight of overtourism. Other places that are not as well known, such as Timbuktu, Mali, stive to maintain fragile mosques and mausoleums, especially as urbanization threatens their infrastructure (UNESCO, 2025). The tradeoff lies in weighing economic growth and improved urban infrastructure against protecting cultural identity, history, and the continuity of traditions for future generations.

## 8.4 Tradeoffs in Global Agreements

Even international frameworks like the Paris Agreement reflect tradeoffs. Countries agreed to limit global warming to 1.5–2 °C (2.7–3.6 °F), but balancing emissions cuts with economic realities has proven difficult (Rajamani, 2016). Many developing nations argue that strict climate targets restrict their ability to industrialize, while wealthier nations push for faster transitions to clean energy (Peters et al., 2017). The principle of "common but differentiated responsibilities" reflects the compromise that all nations share responsibility for climate action, but developed countries are expected to lead due to their historical emissions (UNFCCC, 2015). Still, the gap between ambition and action remains large, with global emissions continuing to rise. Tradeoffs in these negotiations show that sustainability challenges are not only technical but deeply political, requiring trust, equity, and long-term cooperation.

Brazil's Amazon rainforest illustrates how tradeoffs play out at a global scale. The government has encouraged deforestation for cattle ranching and soybean farming, industries that employ millions and generate billions in export revenue (Barona et al., 2010). However, this comes at the expense of the Amazon's ecosystems, which act as a global carbon sink. The Amazon absorbs about 5% of global $CO_2$ emissions annually (UNEP, 2021). Destroying it undermines global climate goals, creates conflict with Indigenous communities, and risks long-term ecological collapse. The tradeoff here is particularly difficult because it involves local versus global priorities. For Brazilian policymakers, agricultural expansion brings immediate economic and political benefits. For the global community, protecting the Amazon is critical for climate stability, biodiversity conservation, and Indigenous rights. This clash underscores the challenge of aligning national development goals with global sustainability objectives (Nepstad et al., 2014).

Another recent and significant tradeoff in global climate agreements centers on climate finance, particularly the struggle to balance mitigation (cutting emissions) with adaptation (helping vulnerable nations cope with climate impacts). Developed nations committed to mobilize $100 billion annually under the UNFCCC climate finance goal. However progress remains uneven, with many developing countries arguing that actual disbursements have lagged (OECD, 2024). However, only about 28% of those funds went toward adaptation, while 60% supported mitigation efforts (OECD, 2024). This imbalance reveals a fundamental tension, that while mitigation earns more political and private-sector support, adaptation is often more urgent for poor countries, yet harder to fund.

The composition of this climate finance also reflects major tradeoffs. A large portion of the funds comes in the form of loans, rather than grants (OECD, 2024). While loans enable large-scale infrastructure projects, they can increase debt burdens in countries that are already economically fragile. Grants, which would be more equitable and less risky for recipient nations, are growing more slowly, even as the absolute amount of grant-equivalent climate finance has doubled since 2016 (OECD, 2024). This financial structure raises ethical questions, such as are donors merely offloading risk onto poorer nations in the name of climate cooperation?

Even with these financial structures in place, trust remains fragile. Developing countries often complain that pledges are vague or delayed, or that reported climate finance does not translate into real, effective support on the ground. On the other hand, donor countries argue that accountability, transparency, and mobilizing private investment are difficult to guarantee. This underscores that tradeoffs in global climate agreements are not only about numbers. They are rooted in power, risk, and political legitimacy.

An additional tax on fossil fuels for some of the wealthiest countries has been proposed to raise $720 billion by the end of the decade to help developing countries, but has not been passed by governments (Taylor, 2024). While lobbyist have been seen as creating barriers to this, the tax would be passed onto the consumer, whom have already been more cautious with spending due to rising costs and inflation.

Ultimately, tradeoffs in international climate policy illustrate that sustainable development cannot be separated from geopolitics. Even when countries agree on ambitious targets, the mechanisms to achieve them bring their own costs, risks, and controversies. Navigating these tradeoffs requires not just technical planning but also trust-building, equity, and long-term institutional commitment.

## 8.5 Determining Priorities

Prioritizing impacts is essential when sustainability goals compete for attention and resources. Instead of assuming all issues require equal effort, decision makers can identify which interventions produce the greatest improvements for people and ecosystems. This ensures that limited resources are directed toward actions with the highest environmental and social returns, making long term sustainability planning more effective and more equitable.

Some of the most pressing concerns when it comes to decision making and tradeoffs include the issues that:

- Agriculture accounts for about 72% of global freshwater withdrawals (FAO, 2023).
- Roughly 29% of global land is experiencing some level of degradation resulting from soil erosion, nutrient depletion, and pollution (IPBES Global Assessment, 2019).
- Emissions from agrifood systems rose by about 9% between 2000 and 2020 (OECD, 2025).
- Livestock production uses about 77% of agricultural land but supplies 18% of global calories and about 37% of global protein (Poore and Nemecek, 2018).
- About 19% of all food available to consumers, equal to more than 1 billion tons, is wasted across the retail, service, and household sectors (UNEP, 2024).

These figures point to areas where targeted interventions can offer broad benefits. Reducing food waste increases efficiency, lowers emissions, and decreases pressure on land and water. Addressing land degradation improves soil function, biodiversity, and long term food security. Shifting consumption patterns toward foods with lower environmental footprints can reduce emissions and resource use while still supporting nutritional needs.

Interventions with multiple co-benefits are especially valuable. Evidence from a cross national retail study shows that foods with lower environmental impact often have lower unit prices, meaning that more sustainable diets can also support affordability and nutrition goals (Martinez et al., 2024). Actions that provide environmental gains without imposing higher costs on consumers or producers are particularly important in regions facing economic or food security constraints.

It is also essential that prioritization frameworks address equity considerations. Some environmental policies create new burdens for communities that already face barriers related to income, land access, or political representation. Including diverse perspectives in decision making helps ensure that sustainability measures do not deepen inequalities, especially in rural areas where livelihoods depend on natural resources.

Prioritizing impacts highlights tradeoffs clearly, supports more strategic decision making, and improves the likelihood that sustainability goals will strengthen both human well-being and ecological integrity. By identifying where the strongest and most multidimensional benefits occur, societies can pursue sustainability pathways that are practical, fair, and grounded in evidence.

## 8.6 Wrap-Up

Tradeoffs are at the heart of sustainability. They force societies to ask difficult questions about which goals to prioritize and at what cost. Every decision in sustainable development involves balancing competing needs, whether it is economic growth versus environmental protection, urban expansion versus cultural preservation, or short-term gains versus long-term resilience. While some tradeoffs can be managed through innovation, cooperation, and stronger governance, others require difficult ethical and policy choices.

Yet even with the best innovation and planning, tradeoffs are unavoidable. Expanding energy access may still create some environmental impacts, urban development may require difficult decisions about heritage sites, and conservation efforts can sometimes conflict with local economic needs. The key is to recognize these tradeoffs openly, assess both immediate and long-term costs and benefits, and implement policies that distribute risks and rewards fairly.

By studying tradeoffs, students, policymakers, and communities develop a more realistic understanding of sustainability. They learn that achieving a sustainable future is not about perfection, but about making informed, responsible choices that balance human needs, environmental protection, and social equity. Ultimately, understanding and managing tradeoffs allows societies to make progress that is both practical and ethical, supporting long-term well-being for people and the planet.

## References


Barona, E., Ramankutty, N., Hyman, G., & Coomes, O. T. (2010). The role of pasture and soybean in deforestation of the Brazilian Amazon. *Environmental Research Letters, 5*(2), 024002. https://doi.org/10.1088/1748-9326/5/2/024002

FAO. (2021). *The state of the world's land and water resources for food and agriculture—Systems at Breaking Point (SOLAW 2021).* Food and Agriculture Organization of the United Nations. https://www.fao.org/3/cb7654en/cb7654en.pdf

FAO (2023). *AQUASTAT main database.* Food and Agriculture Organization. https://www.fao.org/aquastat

IEA. (2021). *Global energy review 2021.* International Energy Agency. https://www.iea.org/reports/global-energy-review-2021

IPBES. (2019). *Global assessment report on biodiversity and ecosystem services.* Intergovernmental Science Policy Platform on Biodiversity and Ecosystem Services. https://www.ipbes.net/global

IUCN. (2016). *Informing decisions on trophy hunting*. International Union for Conservation of Nature. https://www.iucn.org/resources/issues-brief/informing-decisions-trophy-hunting

Koh, L. P., & Wilcove, D. S. (2008). Is oil palm agriculture really destroying tropical biodiversity? *Conservation Letters, 1*(2), 60–64. https://doi.org/10.1111/j.1755-263X.2008.00011.x

Laibson, D. (1997). Golden eggs and hyperbolic discounting. *The Quarterly Journal of Economics, 112*(2), 443–478. https://doi.org/10.1162/003355397555253

Martinez, E. M., Tichenor Blackstone, N., Wilde, P. E., Herforth, A., & Masters, W. A. (2024). *Environmental impacts, nutritional profiles, and retail prices of commonly sold retail food items in 181 countries*. Cornell University. https://arxiv.org/abs/2401.13159

Nepstad, D. C., et al. (2014). Slowing Amazon deforestation through public policy and interventions in beef and soy supply chains. *Science (Washington, D.C.), 344*(6188), 1118–1123. https://doi.org/10.1126/science.1248525

OECD. (2024). *Climate finance provided and mobilised by developed countries in 2013–2022.* OECD Publishing. https://doi.org/10.1787/19150727-en

OECD (2025). *Agricultural policy monitoring and evaluation 2025.* Organisation for Economic Cooperation and Development. https://www.oecd.org/en/publications/agricultural-policy-monitoring-and-evaluation-2025_a80ac398-en/full-report/making-the-most-of-the-trade-and-environment-nexus-in-agriculture_d74b07a7.html

Peters, G. P., et al. (2017). Key indicators to track current progress and future ambition of the Paris agreement. *Nature Climate Change, 7*(2), 118–122. https://doi.org/10.1038/nclimate3202

Poore, J., & Nemecek, T. (2018). Reducing food's environmental impacts through producers and consumers. *Science, 360*, 987–992. https://doi.org/10.1126/science.aaq0216

Rajamani, L. (2016). Ambition and differentiation in the 2015 Paris agreement: Interpretative possibilities and underlying politics. *International and Comparative Law Quarterly, 65*(2), 493–514. https://doi.org/10.1017/S0020589316000130

Ripple, W. J., et al. (2016). Bushmeat hunting and extinction risk to the world's mammals. *Royal Society Open Science, 3*(10), 160498. https://doi.org/10.1098/rsos.160498

Roberts, J. T., & Parks, B. C. (2007). *A climate of injustice: Global inequality, North–South politics, and climate policy.* MIT Press https://mitpress.mit.edu/9780262182614/a-climate-of-injustice/

Taylor, M. (2024, April 29). *Taxing big fossil fuel firms 'could raise $900bn in climate finance by 2030'.* The Guardian. https://www.theguardian.com/environment/2024/apr/29/taxing-big-fossil-fuel-firms-raise-billions-climate-finance

UNEP. (2021). *Making peace with nature: A scientific blueprint to tackle the climate, biodiversity and pollution emergencies.* United Nations Environment Programme. https://www.unep.org/resources/making-peace-nature

UNEP. (2024). *Food waste index report 2024.* United Nations Environment Programme. https://wedocs.unep.org/handle/20.500.11822/45275

UNESCO. (2020). *World heritage and sustainable development.* United Nations Educational, Scientific and Cultural Organization. https://whc.unesco.org/en/sustainabledevelopment/

UNESCO. (2025). *The list: Timbuktu.* United Nations Educational, Scientific and Cultural Organization. https://whc.unesco.org/en/list/119/

UNFCCC. (2015). *Paris Agreement.* United Nations Framework Convention on Climate Change. https://unfccc.int/process-and-meetings/the-paris-agreement/the-paris-agreement

United Nations. (2019). *World population prospects 2019: Highlights.* UN Department of Economic and Social Affairs. https://population.un.org/wpp/Publications/Files/WPP2019_Highlights.pdf

Unruh, G. C. (2000). Understanding carbon lock-in. *Energy Policy, 28*(12), 817–830. https://doi.org/10.1016/S0301-4215(00)00070-7a

World Bank. (2022). *World development report 2022: Finance for an equitable recovery.* World Bank. https://www.worldbank.org/en/publication/wdr2022

# Chapter 9
# Attitudes, Perceptions, and Behavior Change

## 9.1 Why Behavior Change Matters

We know we have important issues that involve us all making changes to ensure future generations have what they need to survive. So why is it so hard to make these changes when we know the consequences of not doing so are dire?

Changing societal behavior to increase sustainability practices is one of the biggest challenges in achieving long-term goals. Even the most well-crafted policies, technologies, and international agreements cannot succeed without people adopting more sustainable habits in their daily lives. From reducing household energy use to supporting sustainable businesses, individual and collective behaviors shape whether sustainability goals can be met. However, people's attitudes (beliefs and evaluations), perceptions (how they interpret environmental risks and solutions), and social norms (what they see others doing) play a powerful role in determining how much sustainable behavior actually occurs (Ajzen, 1991). These are evaluated through the Theory of Planned Behavior, which looks at changing human behavior.

The difficulty of shifting human and societal behavior lies in psychology. At the individual level, creating new habits is already hard. People may intend to reduce waste or cycle to work, but competing priorities, convenience, or lack of resources often prevent follow-through. On a societal scale, the challenge is multiplied: creating new social norms requires widespread adoption across communities, cultures, and economies.

Psychological research shows that humans are more likely to act on short-term rewards (like convenience or cost savings) than long-term benefits (like climate stability) (Kahneman, 2011). This is why policies and campaigns that appeal to immediate personal benefits, such as saving money on energy bills, often achieve better results than those focused only on distant environmental outcomes.

The UN has long acknowledged the role of education and awareness in changing perceptions. For example, SDG 12 (Responsible Consumption and Production)



N. Hugo, *Towards Sustainability*, https://doi.org/10.1007/978-3-032-29339-8_9

highlights the need for awareness campaigns to promote sustainable lifestyles, while SDG 13 (Climate Action) stresses education as a key strategy for climate adaptation and mitigation.

## 9.2 The Theory of Planned Behavior

One of the most influential frameworks for understanding sustainability-related behavior is the **Theory of Planned Behavior**, which was developed by Icek Ajzen in 1991. This theory suggests that a person's behavior is shaped by three factors:

- *Attitudes*: whether a person believes the behavior is positive or beneficial.
- *Subjective norms*: the impact important people or social groups has on the person, in which their approval or disapproval of the behavior has on the person.
- *Perceived behavioral control*: whether the person feels capable of performing the behavior, given resources available and obstacles or barriers to completing the behavior.

For example, an individual may believe that recycling helps the environment (positive attitude), notice that their neighbors recycle regularly (norm), and feel that recycling bins are easily accessible (control). Together, these factors increase the likelihood that the person will recycle consistently.

Energy consumption provides a clear example of how perceptions and attitudes influence sustainability. Household energy-use feedback from in-home displays or direct monitoring has been shown in several randomized studies to reduce residential electricity consumption by around 5% to 15% under certain conditions. However, evidence also shows that these reductions often fade over time, indicating that feedback alone may not sustain long-term behavioral change (Houde et al., 2012; Kelly & Knottenbelt, 2016). Recognizing this, the UN has supported national programs that combine technology and education. For example, in India, the Ujala program distributed more than 360 million LED bulbs, reducing energy demand while educating consumers about cost savings and environmental benefits (UNDP, 2019). By framing the shift as both environmentally responsible and financially smart, the initiative successfully changed household behavior on a large scale.

Jordan, one of the most water-scarce countries in the world, provides a strong case of how behavior change programs can be effective. With UN support, national campaigns emphasized both the scarcity of water and the cultural value of stewardship. Schools taught children water-saving habits, while public messaging highlighted community responsibility. As a result, per capita water use declined, and the government reported increased public awareness of water scarcity (UNDP, 2017). This case shows how combining education, cultural framing, and accessible infrastructure can overcome behavioral barriers.

The UN frequently applies these principles in its outreach campaigns. For instance, the UN Environment Programme's Clean Seas Campaign addresses attitudes by raising awareness about plastic pollution, norms by highlighting global participation, and control by encouraging local policies that make alternatives to single-use plastics available (UNEP, 2018). So why is it still so difficult to reach the milestones set by the SDGs?

## 9.3 Barriers to Sustainable Behavior

Even when people want to act sustainably, barriers often prevent behavior change:

- *Structural barriers:* lack of infrastructure, such as inadequate public transport or recycling facilities.
- *Economic barriers:* sustainable products may cost more upfront, even if they save money later.
- *Information gaps:* people may not know how their actions impact the environment.
- *Cultural barriers:* traditions or social expectations may discourage certain changes.

For example, in many urban areas in developing countries, residents may wish to reduce waste, but lack access to reliable collection services. Similarly, consumers in wealthier countries may support sustainable fashion in principle but continue buying fast fashion due to affordability and trends (Niinimäki et al., 2020).

In addition, long-term studies find that the average energy consumption reduction attributable to feedback or incentives tends to be modest (often ~3–5%) and that behavioral changes may diminish over time unless reinforced by supportive infrastructure, pricing, or policy measures (Jones, 2015; BIEE, 2025). The UN acknowledges these barriers in its 10-Year Framework of Programmes on Sustainable Consumption and Production (10YFP), which works with governments and private sectors to make sustainable lifestyles easier and more accessible globally (United Nations, 2014).

## 9.4 Perceptions of Risk

Perceptions of risk play a critical role in shaping sustainability behaviors and influencing the degree to which individuals and communities engage with environmental issues. When people view climate change as an immediate and personal threat, or as something that directly affects their health, safety, or livelihood, they are far more likely to take action. This might include supporting stricter environmental policies, changing consumption patterns, or adopting sustainable technologies in their daily lives. However, when climate change is perceived as a distant or abstract issue, the motivation to act diminishes. Many individuals conceptualize global warming as a future problem that will primarily affect other regions, generations, or social groups rather than their own (Weber, 2010). They are less likely to act when an issue is a future problem.

Bridging this psychological distance is crucial for fostering engagement. As some regions experience more frequent heatwaves, rising energy costs, or regional flooding, they are able to connect to the idea of broader climate trends. In Germany, interactive public exhibits have allowed citizens to visualize how everyday choices, including transportation or household energy use, contribute to emissions and extreme weather (MuseumNet, 2024). These campaigns not only informed people, but also reframed climate change as a lived, daily experience. When individuals can draw a direct link between their own actions and visible environmental outcomes, their sense of agency grows, making climate action more tangible and urgent.

## 9.5 Social Norms and Collective Action

Social norms exert a profound influence on sustainability because they shape what people perceive as acceptable or expected behavior within their communities. Individuals are more likely to adopt sustainable practices when they observe peers, neighbors, or respected figures engaging in those behaviors. This dynamic reinforces the concept that environmental action is not solely a personal choice, but a social process that gains momentum as collective participation increases. Social norms also interact with other behavioral factors, such as attitudes and perceived control, creating a reinforcing cycle where visible participation encourages further engagement. In relation to the Theory of Planned Behavior, social norms directly influence intentions, highlighting the importance of community and peer dynamics in driving sustainable choices.

Community-based initiatives often leverage local leadership, peer influence, and social networks to accelerate behavior change. In addition to modeling behaviors, programs that publicly recognize or reward participation can further embed sustainability into community identity. This can be done through certifications, local awards, or public acknowledgments. Over time, these practices shift perceptions, making environmentally responsible behavior not only expected but also a source of social pride.

This also connects with cultural perspectives on technology and behavior in international settings. In international business, the success of new systems depends on their alignment with local cultural norms, values, and social structures (Ferraro, 2006). From this perspective, tools like digital feedback systems, smart infrastructure, and resource monitoring technologies are most effective when they reinforce rather than disrupt existing patterns of social interaction and visibility. Durable sustainability is most likely to emerge when technologies are adapted to local cultural contexts and evolve alongside social norms, creating conditions where sustainable practices are both practical and socially reinforced. When technologies fit within cultural contexts, they can strengthen peer influence and community expectations, increasing the likelihood that sustainable behaviors become normalized and sustained over time.

In Asia, Singapore's national "Clean and Green" campaign provides a clear example of how social norms can be cultivated on a large scale. Through a combination of public education, competitions, and community recognition programs, neighborhoods are encouraged to maintain cleanliness, plant greenery, and reduce waste (Clean and Green Singapore, 2025) Awards for exemplary residents and community groups create visible incentives, signaling that sustainable behavior is socially valued. Over time, these efforts have fostered a sense of collective responsibility, transforming environmental stewardship into an expected aspect of civic life. As communities internalize these norms, sustainable behaviors become self-reinforcing, spreading organically and contributing to broader societal transformation.

## 9.6 Education as a Driver of Change

Education remains one of the most transformative tools for changing perceptions, fostering empathy for the planet, and inspiring long-term shifts in behavior. SDG 4 (Quality Education) explicitly calls for integrating education for sustainable development into learning systems, emphasizing not only scientific knowledge but also values, ethics, and skills for sustainable living.

In Kenya, UNESCO's education programs have illustrated the deep ripple effects of educating people and society. By embedding environmental education into primary school curricula, which can be done through lessons on tree planting, water conservation, and sustainable farming. These students then became ambassadors of change within their families, spreading the information beyond the classroom. When they shared what they learned with parents, they helped influence community agricultural practices and local resource management. Over time, this intergenerational learning helped normalize sustainability as both a household and community goal (UNESCO, 2020). The example highlights how children, when properly empowered, can catalyze social transformation.

Beyond the classroom, the growing trend of businesses adopting fair-trade sourcing practices illustrates how education and intentional ethical business decisions together drive sustainable economic systems. Many companies have learned that by educating themselves about the benefits of fair trade, such as stronger supply-chain resilience, better brand trust, and consumer loyalty, they can improve their overall business performance (Fair Trade USA, n.d.; Fairtrade America, n.d.). For instance, brands that become Fair Trade Certified report advantages including improved supply-chain transparency, enhanced reputation, and differentiated products in a values-driven marketplace (Fair Trade Resource Network, 2023). As more companies recognize that ethical sourcing is not only morally sound, but also economically beneficial. Businesses are increasingly pursue fair-trade models, creating positive economic ripple effects globally and enhancing their own competitiveness.

Another way of increasing education on sustainability practices in society is through infographics in public areas on public service announcements through media, including TV, social media and broadcast radio. High traffic areas, such as grocery stores or shopping centers, can be great places to have the information be seen, but these areas tend to have people that are mobile and less likely to stop and look at information. Places where the public is likely to sit, such as transit areas, health centers, libraries or community centers, are more likely to the information be read and absorbed by the public. Providing the information in multiple formats assists with educating the public in a way where it is easily understood and more likely to be implemented. Education fosters critical thinking, empathy, and a sense of stewardship, ensuring that future generations see sustainability not as an obligation but as a shared responsibility.

Successful public service announcement campaigns use psychology, storytelling, and community engagement to shift norms. The UN has run global initiatives such as World Environment Day and World Water Day, which create shared moments of reflection and action. These campaigns highlight relatable issues, from plastic pollution to deforestation, making sustainability part of everyday conversations. For example, the "Think.Eat.Save" campaign, led by UNEP and the FAO, focused on food waste reduction by appealing to cultural values of not wasting food while also showing the environmental and economic benefits. Evaluations showed increased awareness and reductions in household food waste in several countries (FAO, 2014).

## 9.7 Economic Incentives and Behavior Change

While values and awareness are important, practical incentives often determine whether sustainable behaviors are adopted and sustained. People are far more likely to change their habits when those choices align with personal benefits such as saving money, improving comfort, or enhancing convenience. Programs that successfully integrate economic advantages into sustainable practices tend to achieve lasting results because they appeal simultaneously to rational self-interest and broader social well-being.

In Rwanda, for instance, UN initiatives supporting solar mini-grids provided rural households with affordable, reliable electricity. The shift was motivated not just by environmental ideals but by tangible cost savings and the convenience of stable power access (UNDP, 2020). Families that once spent a significant portion of their income on kerosene and batteries could redirect those savings toward education and small business ventures. As local economies grew around these new energy hubs, such as phone-charging stations and refrigerated food stalls, the benefits multiplied and created an economic and social ripple effects. When sustainable options are also the easiest and most affordable, widespread adoption becomes easier obtain.

A similar dynamic can be seen in India's clean cooking initiatives, which demonstrate how socio-cultural and economic factors reinforce one another in advancing sustainability. The Indian government's *Pradhan Mantri Ujjwala Yojana* program, which launched in 2016, provided subsidies and financial support to help low-income households transition from traditional biomass stoves to cleaner liquefied petroleum gas or electric cookers. This shift was not only environmentally beneficial, but also transformative for public health and gender equity. For millions of rural women, the transition reduced exposure to harmful indoor air pollution, which are responsible for respiratory illnesses and premature deaths (World Bank, 2021). The combination of cultural resonance and economic accessibility helped over 90 million households adopt cleaner energy solutions by 2022 (IEA, 2022). This shows that sustainable behavior flourishes when it is supported not only by cost savings but also by social recognition, improved health, and empowerment.

## 9.8 Challenges, Criticisms and Future Directions

Despite progress, large-scale behavior change remains challenging. Critics argue that focusing too much on individual responsibility can distract from the need for systemic reforms (Shove, 2010). For example, encouraging people to recycle while industries continue mass production of plastics may create an imbalance in responsibility. The UN has responded by emphasizing the dual approach, which is that individuals must change behaviors, but governments and corporations must also provide supportive policies, infrastructure, and incentives. This balance recognizes that behavior change cannot occur in isolation, as it requires supportive environments (United Nations, 2021).

Looking forward, behavior change strategies will need to become more inclusive, culturally sensitive, and technology-driven. The UN has called for expanding education for sustainable lifestyles, embedding sustainability into economic systems, and leveraging behavioral science for policymaking (United Nations, 2021). Small design changes can make sustainable choices easier to adapt. For instance, placing plant-based meals more prominently in cafeterias can increase their selection without restricting choice. Digital tools such as apps or web platforms could allow individuals to track and visualize their environmental footprints, receive tailored recommendations, and monitor progress over time, thereby reinforcing sustainable behavior continuously rather than only temporarily (Houde et al., 2012).

Additionally, large-scale public-lighting or energy-efficiency programs demonstrate how system-level initiatives can support individual behavior change. Under the UJALA Scheme in India, more than 36.87 crore LED bulbs have been distributed, saving roughly 47,883 million kWh per year and reducing peak electricity demand by about 9586 MW, leading to substantial reductions in $CO_2$ emissions (Economic Times, 2023). These kinds of structural interventions lower the cost and increase the convenience of sustainable choices, creating enabling environments that make individual behavior change more feasible and meaningful.

Moreover, combining feedback mechanisms with infrastructure improvements can boost effectiveness. A field study offering real-time electricity consumption feedback achieved an average reduction of 5.7% in household electricity use (Houde et al., 2012). This suggests that when individuals are given both information and the means to act, real and measurable change is possible. Future strategies should therefore integrate behavioral insights, supportive infrastructure, accessible technology, and equitable access to resources so that sustainable behavior becomes the default rather than an added burden. Coupled with global education and digital engagement, such strategies may accelerate change at a larger scale (United Nations, 2021).

## 9.9 Wrap-Up

Changing behaviors, perceptions, and social norms is central to advancing sustainability, but incredibly difficult to do. Individual actions, from energy use to waste management, shape whether broader policies and technological innovations achieve their intended impact. Behavioral frameworks, such as the Theory of Planned Behavior, reveal that attitudes, social influences, and perceived control interact to determine whether people adopt sustainable practices. When these elements are aligned, sustainable choices become easier, more consistent, and more likely to spread within communities.

Combining education, cultural framing, infrastructure and resources, and economic incentives can produce meaningful change, but requires understanding the target audience and analyzing how significant the changes are. Initiatives such as LED distribution programs in India, water conservation campaigns in Jordan, solar mini-grids in Rwanda, and clean cooking programs in India show that sustainability is most effective when immediate benefits, social recognition, and long-term environmental goals are integrated. Similarly public service campaigns, community-led projects, and education programs illustrate the importance of social norms and collective engagement in reinforcing sustainable behaviors. When people see their peers acting responsibly and have access to clear guidance and supportive infrastructure, sustainable habits become embedded into daily life.

Economic and social incentives are also critical. Programs that reduce costs, improve convenience, or provide additional social and health benefits encourage wider adoption of sustainable choices. Education and awareness alone may inspire intent, but behavior only changes when actionable pathways are visible and feasible. Leveraging behavioral insights alongside technology, policy, and community engagement ensures that sustainability becomes both practical and culturally resonant.

Sustainability ultimately depends on the interplay between individual action and systemic support. When governments, businesses, and communities collaborate to make sustainable practices accessible, desirable, and normative, behaviors can shift at scale. Embedding sustainability into everyday life transforms it from an ideal into a shared responsibility, demonstrating that lasting change emerges from aligning human motivation with collective environmental and social goals.

## References


Ajzen, I. (1991). The theory of planned behavior. *Organizational Behavior and Human Decision Processes, 50*(2), 179–211. https://doi.org/10.1016/0749-5978(91)90020-T

BIEE. (2025). Incentivising households to reduce electricity consumption: A meta-analysis of the experimental evidence. *Behavioural Insights for Energy Efficiency, 105.*

Clean & Green Singapore. (2025). *About Clean & Green Singapore.* https://www.cgs.gov.sg/

Economic Times. (2023, February 1). *UJALA scheme completes 10 years, saves 19,153 crore annually.* Economic Times https://energy.economictimes.indiatimes.com/news/power/ujala-scheme-completes-10-years-saves-19153-crore-annually/117008100

Fair Trade Resource Network. (2023). *Benefits of fair trade certification: Creating ethical supply chains.*. https://fairtraderesourcenetwork.org/uncategorized/benefits-of-fair-trade-certification-creating-ethical-supply-chains/.

Fair Trade USA. (n.d.). *Our impact: Benefits & advantages of Fair Trade Certified.* https://www.fairtradecertified.org/what-we-do/our-impact/

Fairtrade America. (n.d.). *Key benefits of Fairtrade.* https://www.fairtrade.net/us-en/why-fairtrade/what-we-do/key-benefits.html

FAO. (2014). *Think.Eat.Save: Reduce your foodprint.* Food and Agriculture Organization & United Nations Environment Programme. https://www.unep.org/thinkeatsave

Ferraro, G. P. (2006). *The cultural dimension of international business* (5th ed.). Pearson Prentice Hall. https://openlibrary.org/books/OL17154111M/The_cultural_dimension_of_international_business

Houde, S., Aldy, J. E., & Auffhammer, M. (2012). Real-time feedback and electricity consumption: A field experiment. *Energy Journal, 33*(2), 1–27.

IEA. (2022). *India energy outlook 2022.* International Energy Agency. https://www.iea.org/reports/india-energy-outlook-2022

Jones, A. W. (2015). Perceived barriers and policy solutions in clean energy infrastructure investment. *Journal of Cleaner Production, 104*, 297–304.

Kahneman, D. (2011). *Thinking, fast and slow.* Farrar, Straus and Giroux. https://us.macmillan.com/books/9780374533557/thinkingfastandslow

Kelly, J., & Knottenbelt, W. (2016). Does disaggregated electricity feedback reduce domestic electricity consumption? A systematic review of the literature. *Energy Policy., 94*, 317–327.

MuseumNet. (2024). *How to create an exhibition about the climate crisis.* https://www.museumnext.com/article/klima_x-an-exhibition-on-communicating-the-climate-crisis/#:~:text=As%20a%20curator%2C%20sociologist%2C%20process,become%20part%20of%20the%20solution?%E2%80%9D

Niinimäki, K., Peters, G., Rissanen, T., Gwilt, A., & Karell, E. (2020). The environmental price of fast fashion. *Nature Reviews Earth & Environment, 1*(4), 189–200. https://doi.org/10.1038/s43017-020-0039-9

Shove, E. (2010). Beyond the ABC: Climate change policy and theories of social change. *Environment and Planning A, 42*(6), 1273–1285. https://doi.org/10.1068/a42282

UNDP. (2017). *Water governance in the Arab region.* United Nations Development Programme. https://www.undp.org/arab-states/publications/water-governance-arab-region

UNDP. (2019). *Energy efficiency and renewable energy in India.* United Nations Development Programme. https://www.undp.org/india/projects/energy-efficiency-renewable-energy

UNDP. (2020). *Scaling up energy access through solar mini-grids in Rwanda.* United Nations Development Programme. https://www.undp.org/rwanda/publications/scaling-energy-access-through-solar-mini-grids

UNEP. (2018). *Clean seas campaign.* United Nations Environment Programme. https://www.unep.org/cleanseas

UNESCO. (2020). *Education for sustainable development goals: Learning objectives.* UNESCO. https://unesdoc.unesco.org/ark:/48223/pf0000261800

United Nations. (2014). *10-year framework of Programmes on sustainable consumption and production*. United Nations. https://www.un.org/esa/dsd/csd/csd_pdfs/csd-19/Background-paper-10YFP.pdf

United Nations. (2021). *Our common agenda*. United Nations. https://www.un.org/en/content/common-agenda-report/

Weber, E. U. (2010). What shapes perceptions of climate change? *Wiley Interdisciplinary Reviews: Climate Change, 1*(3), 332–342. https://doi.org/10.1002/wcc.41

World Bank. (2021). *The benefits of clean cooking in India: Impacts of the Pradhan Mantri Ujjwala Yojana (PMUY) Programme*. https://www.worldbank.org/en/topic/energy/publication/the-benefits-of-clean-cooking-in-india

# Chapter 10
# Emerging Policy and Trends in Sustainability

## 10.1 Why Policy Change Matters

Global policies and emerging trends for sustainability are reshaping how nations, businesses, and individuals approach environmental, economic, and social challenges. As the world faces escalating crises such as climate change, biodiversity loss, and resource depletion, the need for more sustainable practices has become urgent. Governments and organizations increasingly recognize that piecemeal reforms are insufficient, as systemic transformations are required. Policies that integrate long-term thinking, accountability, and multi-sector collaboration are necessary to address complex global challenges that span borders and sectors.

As we have seen in previous chapters, the United Nations has played a central role in guiding global policy through frameworks like the 2030 Agenda for Sustainable Development and the Paris Agreement. These platforms emphasize shared responsibility and cooperation, while also encouraging the development of innovative approaches such as circular economies, green finance, and life cycle-based policy analysis (United Nations, 2015). By incorporating these strategies into policy frameworks, governments can influence production, consumption, and investment decisions that collectively reduce environmental impact while promoting social and economic well-being.

Policy change also matters because it drives the adoption of best practices and creates incentives for compliance. Regulations on emissions, renewable energy targets, and resource efficiency standards not only hold organizations accountable but also signal market expectations that influence investment and innovation. Businesses are increasingly responding to these signals by integrating sustainability into corporate strategy, designing products for durability and recyclability, and measuring their environmental footprint throughout the supply chain. Life cycle-based policy analysis, in particular, allows decision makers to consider the full range of



N. Hugo, *Towards Sustainability*, https://doi.org/10.1007/978-3-032-29339-8_10

environmental, social, and economic impacts of a product or activity, helping avoid unintended consequences and encouraging more holistic solutions.

Furthermore, policy change shapes societal norms. As governments introduce sustainability-focused legislation, education, and public campaigns, citizens and communities become more aware of their role in driving change. Policies that promote public transportation, waste reduction, and energy efficiency encourage behaviors that collectively reduce ecological impact. This cultural shift, combined with systemic policy measures, can accelerate the transition toward more sustainable economies and resilient communities.

In essence, understanding policy change highlights the interconnectedness of governance, markets, and society. It demonstrates that sustainability is not only about individual actions but also about creating frameworks that guide collective behavior toward long-term ecological, social, and economic stability. Observing these trends helps forecast how businesses and urban planning strategies will evolve in response to global priorities, ensuring that sustainable development becomes a central, rather than optional, component of future decision making.

## 10.2 The Shift toward Systems Thinking

With the development of the SDGs, we have seen more of a focus on creating synergies between sustainable initiatives. Traditional sustainability policies often addressed single issues, such as pollution control or energy efficiency, one at a time. However, sustainability challenges moving forward are focusing on interconnectedness. For example, improving agricultural output to reduce hunger (SDG 2) can also affect water consumption (SDG 6), biodiversity (SDG 15), and climate resilience (SDG 13).

This recognition has led to the rise of systems thinking in policy, where strategies consider entire networks of social, environmental, and economic impacts. Tools like **Life Cycle Analysis (LCA)** have become central to this approach, providing a holistic view of a product or service's environmental footprint from raw material extraction to disposal (Guinée et al., 2011). Essentially, a LCA looks at materials used, the assembly process, overall use of the product and disposal of a product. A product that can be repurposed after it has finished its usable life for its original purpose is looked at more positively than one that ends up in a landfill at the end of its life.

The One Planet Network, hosted by UNEP, is a global multi-stakeholder platform that promotes sustainable consumption and production and mobilizes governments, businesses, civil society, and researchers to implement the 10-Year Framework of Programmes (10YFP). It is a UN-endorsed commitment to sustainable consumption and production under SDG 12. (UNEP, 2021). Through its six core programs, which include Public Procurement, Tourism, Consumer Information, Buildings and Construction, Food Systems, and Lifestyles and Education, the network provides technical guidance and facilitates partnerships that embed sustainability across entire value chains. By connecting diverse stakeholders and aligning their goals, the One Planet Network helps countries and industries implement systemic changes that promote circular economies, responsible production, and equitable global development.

Beyond UN initiatives, businesses have been more mindful about including LCA in their general practices. LCA may not only improve resource efficiency, but also enhance brand reputation, employee engagement, and customer loyalty in markets where sustainability is valued. Construction projects have included looking at sourced materials, as well as their waste produced during the building process. The fashion industry is looking at alternative materials for their clothing, such as bamboo and hemp, as they grow faster and with less water than traditional materials, including cotton. Some fashion brands are also focusing on durability, as clothing waste significantly contributes to leaching toxic chemical and dyes into the groundwater and soil when they are disposed of in landfills. Pharmaceutical companies are also utilizing LCA methods, to analyze their energy consumption and chemical applications. An emphasis has been also placed on proper disposal of medications as pouring them into the sink or toilet has led to issues with water contamination, which has been difficult to remedy. These examples show how LCAs have been incorporated into businesses more and more throughout recent years, but it has also impacted policy creation as well.

## 10.3 Life Cycle Analysis in Policy

As the LCA is an assessment method that evaluates the total environmental impacts of a product or service, we can think about what areas are in most need of regulations and would benefit from policies to ensure it is implemented where it is needed most. By examining resource use, emissions, and waste across the full life cycle, policymakers are provided evidence-based insights to design regulations. This includes examining extraction, production, distribution, use, and disposal.

For example, the European Union uses LCA in its Ecodesign Directive, which influences everything from energy efficiency standards for appliances to recycling policies for electronics. The UN has promoted similar approaches through the Life Cycle Initiative, a partnership between UNEP and the Society of Environmental Toxicology and Chemistry (SETAC). This initiative helps developing countries adopt LCA for decision-making, ensuring sustainability policies are grounded in data rather than assumptions (UNEP, 2020).

While LCA helps identify problems, **Cradle-to-Cradle** design offers solutions by rethinking products as part of a regenerative system. Unlike traditional "cradle-to-grave" models that end with disposal, Cradle-to-Cradle seeks to create products that can be fully reused, recycled, or biodegraded, minimizing waste. The UN has supported closed-loop and circular production approaches through the 10 Year Framework of Programmes on Sustainable Consumption and Production Patterns framework. Many industries, including textiles, are being encouraged under this program to shift toward circular product design and materials reuse (UNEP, 2025).

LCA has been influential in the emerging trend of striving for a circular economy. Unlike the linear "take-make-dispose" model, the circular economy emphasizes designing out waste, keeping products in use, and regenerating natural systems (Ellen MacArthur Foundation, 2015).

Proponents argue that circular economy and life-cycle thinking support not only SDG 12 (Responsible Consumption and Production), but can also contribute to SDG 8 (Decent Work and Economic Growth), SDG 9 (Industry, Innovation and Infrastructure), SDG 13 (Climate Action), and SDG 15 (Life on Land), by promoting resource efficiency, reducing waste, and encouraging sustainable industrial and consumption patterns (UNEP, 2021; van Buren et al., 2023). Unlike traditional linear production models, which follow a "take-make-dispose" pattern, circular systems emphasize resource efficiency, designing out waste, and regenerating natural systems (Ellen MacArthur Foundation, 2015).

Recognizing its importance, UNEP launched the Global Alliance on Circular Economy and Resource Efficiency (GACERE) in 2021, in collaboration with the European Commission and other partners. GACERE serves as a platform to align governments, businesses, and civil society organizations in scaling up circular practices worldwide (UNEP, 2021). By connecting countries from the Global North and South, GACERE fosters knowledge-sharing and capacity-building to ensure that circular policies can be adapted to different economic and social contexts. For instance, GACERE facilitates technical support for developing countries that may lack infrastructure for recycling or advanced waste management but still face high levels of resource pressure.

## 10.4 ISO 14001, Environmental Management Systems and Sustainable Certifications

Emerging sustainability trends include global standards that harmonize practices across industries. The **ISO 14001** environmental management standard provides a framework for organizations to systematically improve environmental performance. These are voluntary standards in the environmental field that help to guide companies around the world on acceptable practices to minimize negative impacts to the environment and prevent an increase in pollution. By adopting ISO 14001, companies commit to reducing environmental risks, complying with regulations, and continuously improving practices (ISO, 2015).

UN agencies frequently promote ISO standards as part of capacity-building. For example, UNIDO (United Nations Industrial Development Organization) supports small and medium-sized enterprises (SMEs) in Africa to adopt ISO 14001, enabling them to access global markets while improving sustainability performance. An example is the support it provided in Ethiopia to expand its agro-industrial parks, with $1.2 billion being committed for public and private scale ups, which will assist with increasing employment opportunities and reducing poverty in rural areas (UNIDO, 2022).

Other Environmental Management Systems and sustainability certifications include:

- *RC14001 (Responsible Care 14,001)*
  A technical specification created by the American Chemistry Council (ACC) that combines the international ISO 14001 environmental management system requirements with the ACC's Responsible Care principles (which add health, safety, security and community engagement). It lets chemical companies, and other firms that choose to adopt it, be audited and certified against a management system that covers environmental, health, and safety performance (American Chemistry Council, 2023).
- *Planet Mark*
  A UK-based sustainability certification that verifies an organization's carbon footprint and wider environmental and social progress. Planet Mark supports businesses by measuring emissions, setting reduction targets, and reporting progress. It is commonly used by companies and development projects (Planet Mark, 2024).
- *Green Mark (UK)*
  A UK certification and online Environmental Management System designed to help businesses measure, manage, and reduce environmental impacts (energy, waste, water, etc.). It provides a practical route to improve environmental performance and demonstrate credibility to customers and stakeholders (Green Mark, 2025).
- *LEED (Leadership in Energy and Environmental Design)*
  LEED is a globally recognized green building certification developed by the U.S. Green Building Council (USGBC). It provides a framework for designing, constructing, operating, and maintaining buildings that are environmentally responsible, energy-efficient, and healthy for occupants. LEED certification uses a point system that evaluates areas such as energy efficiency, water conservation, waste reduction, and sustainable materials (U.S. Green Building Council, 2025).
- *BREEAM (Building Research Establishment Environmental Assessment Method)*
  A leading international sustainability assessment method for buildings and infrastructure, originally developed in the UK. BREEAM evaluates projects across categories such as energy, water, materials, waste, ecology, resilience, and health and wellbeing. Certification levels (Pass, Good, Very Good, Excellent, and Outstanding) indicate increasing performance. It is one of the most widely used green building certifications globally and is backed by extensive technical standards (BRE, 2025).

As these systems continue to be implemented into various industries, improvements and additional recommendations for better practices continue to be evaluated and implemented. With many different areas seeing the importance of standardized practices to help their industry to sustain performance, it is likely that more Environmental Management Systems will emerge in the future.

## 10.5 Critiques, Challenges and the Future

Emerging sustainability policies and trends offer significant promise, but they face practical challenges and criticisms. A central concern revolves around the reliability and accountability of reported metrics. Environmental claims can sometimes be exaggerated or misleading to attract investors or enhance public image. Even with standardized benchmarks or regulatory oversight, sustainability initiatives risk becoming marketing tools rather than genuine instruments of social or environmental change. This raises important questions about how sustainability performance can be meaningfully measured, verified, and compared across industries and borders.

Another challenge lies in the uneven global implementation of circular economy practices. While initiatives such as recycling, resource efficiency, and waste reduction aim to minimize environmental impact, they tend to favor countries with strong infrastructure, governance, and technological capacity (Hickel, 2020). Developing nations may lack access to financing, technical expertise, or institutional support, limiting their ability to adopt similar strategies. Without targeted interventions, this imbalance risks widening the global sustainability divide, allowing wealthier economies to benefit from circular innovation while poorer nations continue to bear disproportionate environmental burdens. International organizations have sought to address these disparities through capacity-building programs, knowledge-sharing initiatives, and financial mechanisms. The United Nations, for instance, emphasizes inclusive strategies that combine technology transfer with financial support, as reflected in SDG 17, which advocates global partnerships to mobilize resources and share solutions equitably (United Nations, 2021).

A related issue is the potential for overregulation, particularly for businesses. Stricter reporting requirements, operational restrictions, and environmental standards can create significant compliance costs and administrative burdens, especially for smaller firms. Overly rigid policies may favor large corporations capable of navigating complex legal and technical frameworks while limiting opportunities for startups and smaller enterprises. Examples include the European Union's Corporate Sustainability Reporting Directive (CSRD), which requires extensive reporting across more than 1000 environmental and social data points (European Parliament, 2022), and the U.S. Securities and Exchange Commission's climate-risk disclosure rule, which is estimated to cost large firms $640,000 in the first year and $530,000 annually thereafter, with smaller firms also facing high expense burdens (Rajgopal, 2022; SEC, 2024). While regulation is essential for transparency and accountability, excessive rigidity may stifle innovation, reduce agility, and concentrate market power. Striking a balance between oversight and flexibility is therefore critical.

Looking forward, sustainability policy is likely to become more integrated, combining diverse tools under coherent frameworks. Policies may increasingly link Life Cycle Assessment (LCA) requirements with circular economy incentives, supported by ISO standards and financed through mechanisms such as green bonds or sustainable investment funds. Digital technologies, including blockchain-based traceability, AI-driven resource optimization, and data transparency platforms, can enhance both accountability and efficiency (OECD, 2022). The United Nations' *Our Common Agenda* calls for "turbocharged multilateralism," emphasizing collaboration between science, finance, technology, and civil society to accelerate sustainability outcomes (United Nations, 2021). Ultimately, the effectiveness of future sustainability governance will depend on balancing top-down regulation with bottom-up innovation, ensuring policies are both rigorous and adaptable, equitable, and inclusive.

## 10.6 Wrap-Up

Emerging policies and trends in sustainability, including LCA, Cradle-to-Cradle design, ISO 14001 standards, and circular economy models, are reshaping how societies address environmental, economic, and social challenges. These approaches provide practical frameworks for managing resource use, waste, emissions, and social impacts across the full life cycle of products and services. By embedding these strategies into public policy, business practices, and community initiatives, sustainability is becoming more systematic, data-driven, and integrated into everyday decision making. Global frameworks, such as the United Nations' 2030 Agenda for Sustainable Development and the Paris Agreement, have encouraged cooperation between nations, industries, and communities while promoting shared responsibility and long-term planning (United Nations, 2015; UNEP, 2021).

Standards and certifications, including ISO 14001, LEED, Planet Mark, and RC14001, help organizations measure performance, set targets, and continuously improve environmental and social outcomes. Life cycle thinking and circular economy principles complement these standards by promoting regenerative practices, reducing waste, and preventing unintended consequences that can arise when problems are solved in isolation. Policies that integrate these approaches encourage transparency, accountability, and innovation across sectors while highlighting the interconnectedness of environmental, social, and economic systems.

At the same time, challenges remain. Overregulation, uneven access to technology, and varying institutional capacity can create barriers for smaller businesses and developing nations, highlighting the need for flexible, equitable, and adaptive policy frameworks. Efforts to provide technical assistance, capacity building, and financial support can help address these gaps, ensuring that sustainable practices are feasible and impactful across diverse contexts.

These trends illustrate that meaningful sustainability requires coordination across governance, business, and society. When policies are implemented thoughtfully, they can drive measurable improvements, support resilient communities, and foster development that balances human needs with environmental protection and long-term social equity.

## References


American Chemistry Council. (2023). *RC 14001: 2023 technical specification.* https://store.americanchemistry.com/products/rc14001-23

BRE. (2025). *BREEAM: The world's leading science-based suite of validation and certification systems for a sustainable built environment.* https://bregroup.com/products/breeam/

Ellen MacArthur Foundation. (2015). *Towards the circular economy.* EMF. https://content.ellenmacarthurfoundation.org/m/27265af68f11ef30/original/Towards-the-circular-economy-Vol-1.pdf

European Parliament. (2022). *Corporate sustainability reporting directive (CSRD).* European Parliament. https://www.europarl.europa.eu/topics/en/article/20221110STO53001/corporate-sustainability-what-the-eu-expects-from-companies

Green Mark. (2025). *Green Mark certification.* https://greenmark.co.uk/

Guinée, J. B., Heijungs, R., & Huppes, G. (2011). Life cycle assessment: Past, present, and future. *Environmental Science & Technology, 45*(1), 90–96. https://doi.org/10.1021/es101316v

Hickel, J. (2020). *Less is more: How degrowth will save the world.* Penguin.

ISO. (2015). *ISO 14001: 2015.* https://www.iso.org/standard/60857.html

OECD. (2022). *Blockchain for good: Using distributed ledger technologies for sustainability.* Organisation for Economic Co-operation and Development. https://www.oecd.org/digital/blockchain/

Planet Mark. (2024). *About planet Mark.* https://www.planetmark.com/

Rajgopal, S. (2022). *The potential costs of the SEC's climate disclosure rule for companies.* Harvard Law School Forum on Corporate Governance. https://corpgov.law.harvard.edu/2022/08/22/the-proposed-sec-climate-disclosure-rule-a-comment-from-shivaram-rajgopal/

SEC. (2024). *SEC adopts rules to enhance and standardize climate-related disclosures for investors.* U.S. Securities and Exchange Commission. https://www.sec.gov/news/press-release/2024-37

United Nations. (2015). *Transforming our world: The 2030 agenda for sustainable development.* United Nations. https://sdgs.un.org/2030agenda

United Nations. (2021). *Our common agenda.* United Nations. https://www.un.org/en/common-agenda

UNEP. (2020). *Life cycle initiative.* United Nations Environment Programme. https://www.unep.org/topics/finance-and-economic-transformations/driving-transformation-science-and-knowledge/life-cycle

UNEP. (2021). *Catalysing science-based policy action on sustainable consumption and production.* https://www.oneplanetnetwork.org/sites/default/files/report_unea5_catalysing_science-based_policy_action_on_scp_-_task_group_irp-one_planet_0.pdf

UNEP. (2025). *10 year framework of programmes on sustainable consumption and production patterns (10YFP).* https://www.unep.org/10yfp-10-year-framework-programmes-sustainable-consumption-and-production-patterns

UNIDO. (2022). *Annual report 2021.* United Nations Industrial Development Organization. www.unido.org/sites/default/files/unido-publications/2023-02/UNIDO_AR2021_EN.pdf

U.S. Green Building Council. (2025). *What is LEED?* https://www.usgbc.org/leed

van Buren, N., Demmers, M., van der Heijden, R., & Witlox, F. (2023). How can the circular economy support the advancement of the sustainable development goals (SDGs)? A comprehensive analysis. *Sustainability and Production & Consumption, 40*, 352–362. https://doi.org/10.1016/j.spc.2023.07.003

# Chapter 11
# Marketing and Certifications

## 11.1 What Is Greenwashing?

As sustainability becomes a central concern for businesses and consumers, companies are increasingly eager to highlight the environmental and social benefits of their products. Labels such as *eco-friendly*, *sustainable*, *green*, or *carbon-neutral* are now common across industries ranging from fashion to food. However, not all of these claims hold up to scrutiny. A significant problem in this space is **greenwashing**, which refers to the practice of making misleading or exaggerated claims about a product's environmental benefits in order to attract environmentally conscious consumers (Delmas & Burbano, 2011). Greenwashing can take many forms, including vague language, unverified certifications, or selective disclosure of information.

However, legitimate sustainability certifications do exist, and these play a vital role in helping consumers distinguish between genuine efforts and marketing gimmicks. Understanding the difference between credible certifications and misleading claims is essential to ensuring that consumer choices truly support sustainability. This can be difficult for consumers to determine because some companies use logos that look genuine, but have no real backing to them. Therefore consumers must educate themselves on which ones are credible before purchasing, making the process more cumbersome.

## 11.2 Greenwashing in Practice

Greenwashing undermines trust in sustainability marketing and can discourage consumers from engaging in environmentally friendly behaviors. Some common strategies include:



N. Hugo, *Towards Sustainability*, https://doi.org/10.1007/978-3-032-29339-8_11

- *Vague or meaningless language*: Phrases like "all-natural" or "eco-safe" without scientific backing or clear definitions.
- *Hidden trade-offs*: Highlighting one small positive feature (such as packaging made from recycled material) while ignoring larger negative impacts (such as the product's overall carbon footprint).
- *Irrelevant claims*: Promoting a product as free from certain harmful substances (such as CFCs in aerosols), even though such substances are already banned by law.
- *Imagery and branding*: Using green colors, trees, or earth symbols to suggest sustainability without substantive action.
- *Unverified or misleading certifications*: Using a label or tick mark that appears credible but is self-declared or not independently audited.

An example of greenwashing occurred in the Volkswagen emissions scandal, when the company marketed its diesel cars as environmentally friendly and low-emission, while secretly installing devices to cheat emissions tests. This deception not only misled millions of consumers but also caused severe reputational damage and legal penalties for the company (Hotten, 2015). Other forms of greenwashing do not need to be so severely fraudulent to still mislead costumers about products.

In the fashion industry, more subtle forms of greenwashing occurred through exaggerated claims. Many fast fashion brands claim to have "conscious" or "sustainable" collections. Yet investigations show that most of these products are made with the same environmentally harmful practices as their regular lines, with only minor adjustments like slightly higher percentages of recycled polyester (Changing Markets Foundation, 2021). Marketing products to be better than what they are is not new to advertising, but with many people skeptical about the quality or effectiveness of sustainable products, the marketing efforts must be genuine in order to gain the trust of the public. The rise of certifications in sustainability has assisted with this, but some logos that claim that a product is certified do not actually have a process to evaluate or authenticate a product. Therefore understanding which certifications have a set of standards and use a verification process is needed.

## 11.3 The Role of Certifications

As sustainability certifications have been developed to give consumers reliable information about products and services, it is important to understand which ones work towards improving sustainability practices for businesses and which ones are still just greenwashing. These certifications are often awarded by independent organizations and require companies to meet specific standards.

Some of the most recognized certifications include:

- *Fairtrade International*: Ensures fair wages and safe working conditions for farmers and workers, while promoting sustainable farming practices.
- *Forest Stewardship Council (FSC)*: Certifies products sourced from responsibly managed forests that provide environmental, social, and economic benefits.

- *Rainforest Alliance*: Promotes biodiversity conservation and sustainable livelihoods by certifying agricultural and forestry products.
- *LEED (Leadership in Energy and Environmental Design)*: An international certification system that evaluates the sustainability of buildings and infrastructure projects.
- *ISO 14001*: A standard developed by the International Organization for Standardization that provides companies with guidelines for effective environmental management systems.

These certifications can help consumers make better choices, but they are not immune to criticism. Some argue that voluntary certification schemes can be costly for small producers and may exclude marginalized communities, while others note that even certified companies sometimes fall short of full sustainability. It is important to understand that none of these certifications are flawless, but they are taking steps in the right direction to help create standards for evaluation.

## 11.4 United Nations and Certification Initiatives

The United Nations Environment Programme (UNEP) has highlighted the importance of sustainability certifications in guiding consumer behavior and is working towards the improvement of these achieve higher levels of sustainability and improve consumer confidence in buying. The UN has promoted eco-labeling as a tool to encourage sustainable production and consumption, particularly in line with SDG 12 (Responsible Consumption and Production) (UNEP, 2019). In its communications it notes that within the EU "53 % of green claims give vague, misleading or unfounded information" and "40 % of claims have no supporting evidence" (European Commission, 2023). This is very concerning and shows there is a long way to go to improve standardization and certification processes, as well as monitoring for businesses that participate in greenwashing.

The UNEP has supported the development of national eco-labeling systems in countries such as China, India, and Brazil, where consumer markets are rapidly expanding. By building trust in labels, these systems help shift demand toward more sustainable products while discouraging misleading claims. Additionally, the United Nations Global Compact, the world's largest corporate sustainability initiative, encourages companies to align their business strategies with universally accepted principles in human rights, labor, environment, and anti-corruption. While not a certification system per se, it provides reporting mechanisms that increase accountability (UN Global Compact, 2022).

The European Commission has one of the more advanced regulatory frameworks in the world for addressing greenwashing, combining eco-labels with evolving regulation. The EU Ecolabel was first established in 1992 to certify products that meet high environmental standards across their lifecycle, from production to disposal (European Commission, 2023). More recently the Commission published a proposal in March 2023 for a EU **Green Claims Directive** that would require companies to provide scientific evidence for environmental claims made in marketing, and ensure that labels and claims are reliable, comparable and verifiable across the EU. However, as of 2025, the directive remains under negotiation, and some sources indicate the proposal may be withdrawn or significantly delayed (Carbon Trust, 2025). Even if this does pass, it is not an international agreement and would only impact marketing in the EU. This highlights the evolving and partly uncertain nature of the regulatory environment.

## 11.5 Consumer Awareness and Behavior

Ultimately, combating greenwashing and promoting credible certifications also depends on consumer education and awareness. A recent cross-national survey by the Bureau Européen des Unions de Consommateurs (BEUC) and International Consumer Research & Testing (ICRT), conducted in 16 European and non-European countries in 2023, provides important insights into how consumers perceive and respond to green claims (BEUC & ICRT, 2023).

According to the survey:

- 34% of respondents reported having noticed greenwashing at least once in the past 12 months.
- Only 3% said they could *always* distinguish false or misleading green claims.
- 48% of respondents indicated a preference for purchasing products bearing an environmental label compared to those without such a label.

These figures indicate that consumers are show increasing interest and even a preference for sustainable labelled products. However, many feel ill-equipped to accurately evaluate the veracity of environmental claims. In other words, consumer demand for "green" is rising, but consumer capacity to discern genuine claims from marketing spin remains limited.

This gap has multiple implications. With only 3% feel confident in distinguishing misleading claims suggests that the majority of consumers operate under significant uncertainty. That uncertainty may lead to decision-avoidance, where consumers refrain from making green purchases because of doubt, or worse, to inadvertent support of products that are greenwashed.

The preference for labelled products (48%) indicates that certifications or labels still exert strong influence. While this is a positive signal for sustainability marketing, this influence may be undermined if consumers cannot trust or verify them. In effect, labels alone are necessary but not sufficient as they must be accompanied by clarity, transparency, and consumer education. Given the prevalence of greenwashing (as perceived by roughly one-third of the sample), there is a risk of erosion of consumer trust. If consumers frequently encounter misleading claims, their confidence in sustainability labels, brands, or even certifications may decline, leading to more skepticism, reduced purchasing of sustainable products, or backlash.

Therefore, to strengthen consumer behavior in favor of sustainable choices several actions are needed, which include:

- *Transparency:* Companies must provide clear, verifiable evidence of environmental claims, preferably via third-party audited certifications or publicly accessible verification data.
- *Clarity of claims for less consumer literacy:* Educational efforts are required so that consumers can better understand environmental claims (such as "organic" vs "made from organic ingredients"). The Organisation for Economic Co-operation and Development (OECD) underscores that consumers often struggle to assess environmental attributes because they are less directly observable than other product features (OECD, 2024). Reducing the need for consumer education may impact consumer confidence when buying.
- *Regulatory support & enforcement*: Public authorities should support label verification, enforce anti-greenwashing rules, and ensure that environmental claims are subject to scrutiny. Consumer desire for verification is strong as three quarters of respondents in the BEUC/ICRT survey indicated that green claims should only be authorized if pre-approved or verified (BEUC & ICRT, 2023).
- *Empowering consumer choice*: Enabling consumers to make informed sustainable consumption decisions means simplifying environmental information, reducing label overload, and improving trust in high-integrity labels. Label fatigue (too many logos with varying rigor) is itself a barrier. The survey found that over half (56%) of respondents believed there were too many environmental labels and logos (BEUC & ICRT, 2023).

Although consumer interest in sustainable purchasing is increasing, the ability of consumers to navigate the green claims environment remains weak. In that context, improved transparency, stronger regulation, robust certifications, and consumer education are all critical components of creating a marketplace where green claims can genuinely fulfil their promise rather than simply contributing to confusion or skepticism.

## 11.6 Wrap-Up

Marketing and certifications play a pivotal role in the global sustainability movement. However, the modern sustainability marketplace exists *at a crossroads between genuine transformation and strategic illusion*. Greenwashing, which is the act of misleading consumers about the environmental practices or benefits of a company, product, or service, poses a serious challenge to consumer trust, corporate accountability, and the credibility of sustainability marketing. While certifications and eco-labels were originally designed to bridge the information gap between producers and consumers, their rapid proliferation and inconsistent verification processes have, in some cases, contributed to further confusion rather than clarity.

To move toward genuine sustainability, stakeholders must address the complex interplay between corporate behavior, consumer perception, and regulatory frameworks. From a business standpoint, transparency and third-party verification are now strategic necessities. Brands that engage in greenwashing not only risk reputational damage, as evidenced by the Volkswagen emissions scandal, but also face growing legal consequences under evolving regulations like the European Union's proposed Green Claims Directive. The long-term business case for authentic sustainability therefore lies in credibility, consistency, and alignment with recognized standards rather than superficial marketing appeals.

Regulatory efforts at both the national and international levels are increasingly addressing this imbalance. However, regulation and certification alone cannot solve the problem. Consumer awareness and education remain the final and perhaps most powerful defense against greenwashing. This lack of consumer awareness leaves significant space for manipulation through vague or misleading marketing. Therefore, improving environmental literacy must be prioritized through public education campaigns, curriculum integration, and transparent communication from both companies and regulators. Empowering consumers with the skills to critically evaluate green claims helps to create demand for credible certifications and penalize deceptive practices through market dynamics.

In the long term, overcoming greenwashing is not only about policing deceptive claims. It is about transforming the relationship between consumption, trust, and accountability. The credibility of sustainability marketing depends on a shared understanding that authenticity, transparency, and ethical integrity are not marketing trends but prerequisites for a viable future. As the global economy moves toward decarbonization and circularity, credible certifications and informed consumer behavior will be essential in steering industries toward genuine, measurable sustainability outcomes.

## References

Bureau Européen des Unions de Consommateurs & International Consumer Research & Testing. (2023, November 28). The Great Green Maze: How environmental advertising confuses consumers. https://www.beuc.eu/sites/default/files/publications/BEUC-X-2023-149_The_Great_Green_Maze_How_environmental_advertising_confuses_consumers.pdf

Carbon Trust. (2025). *What's next for green claims in Europe and beyond?* A Q&A. https://www.carbontrust.com/news-and-insights/insights/whats-next-for-green-claims-in-europe-and-beyond-a-qa

Changing Markets Foundation. (2021). *Licence to greenwash: How certification schemes are fuelling fossil fashion*. https://changingmarkets.org/report/licence-to-greenwash-how-certification-schemes-and-voluntary-initiatives-are-fuelling-fossil-fashion/

Delmas, M. A., & Burbano, V. C. (2011). The drivers of greenwashing. *California Management Review, 54*(1), 64–87.

European Commission. (2023). *Green claims directive: Ensuring reliable environmental claims*. https://ec.europa.eu/commission/presscorner/detail/en/qanda_23_1693

Hotten, R. (2015, December 10). Volkswagen: The scandal explained. *BBC News*. https://www.bbc.com/news/business-34324772.

OECD. (2024). *Protecting and empowering consumers in the green transition*. Organisation for Economic Co-operation and Development. https://www.oecd.org/content/dam/oecd/en/publications/reports/2025/05/protecting-and-empowering-consumers-in-the-green-transition_c23ea601/12f28e4f-en.pdf

UN Global Compact. (2022). *Annual report 2022: Delivering impact through principles and partnerships*. United Nations. https://ungc-communications-assets.s3.amazonaws.com/docs/about_the_gc/2022%20Annual%20Report_%20Executive%20Summary%20%28English%29.pdf

UNEP. (2019). *Sustainable consumption and production: A handbook for policymakers*. United Nations Environment Programme. https://www.unep.org/resources/report/sustainable-consumption-and-production-handbook-policymakers

# Chapter 12
# Corporate Social Responsibility (CSR)

## 12.1 What Is Corporate Social Responsibility?

Corporate Social Responsibility (CSR) refers to a business approach in which companies incorporate social, environmental, and ethical concerns into their operations, beyond merely making a profit (World Bank, 2020a). Rather than treating CSR as optional philanthropy, modern CSR involves strategic commitments focused on protecting human rights, reducing environmental harm, promoting equitable labor practices, and ensuring ethical governance. It represents a shift in corporate behavior: from transactional charity to long-term responsibility.

The **United Nations Global Compact** defines CSR through its Ten Principles, covering human rights, labor standards, environmental protection, and anti-corruption (UN Global Compact, 2022). By endorsing these principles, businesses signal that they intend to operate in a way that respects both people and the planet. CSR today it is central to business strategy, driven by pressures from consumers, civil society, investors, and governments.

A major driver of CSR is conscious consumerism. With growing awareness of global challenges, such as climate change, income inequality, supply chain labor abuses, and environmental degradation many consumers now demand that companies act responsibly. The World Bank argues that businesses have a crucial role in achieving the United Nations Sustainable Development Goals (SDGs), and that CSR is one vehicle by which the private sector contributes to sustainable development (World Bank, 2020a).



N. Hugo, *Towards Sustainability*, https://doi.org/10.1007/978-3-032-29339-8_12

## 12.2 Historical Evolution of CSR

The notion of CSR has its roots well before the modern sustainability era with companies in the early twentieth century often saw their contribution to society largely through the lens of philanthropy. Wealthy industrialists, such as Andrew Carnegie and John D. Rockefeller, began donating large sums to community causes, cultural institutions, education, and healthcare (Association of Corporate Citizenship Professionals, 2023). This philanthropic CSR was largely paternalistic, rooted in a belief that business elites had a duty to uplift society, though it remained separate from core business operations (Satell Institute, 2023).

In the mid-twentieth century, CSR began to shift from charitable giving to a more formalized concept. A watershed moment occurred in 1953, when economist Howard R. Bowen published *Social Responsibilities of the Businessman*, commonly credited as the first systematic articulation of corporate obligations beyond profit (Satell Institute, 2023). Bowen posed fundamental questions about what social responsibilities businesses should assume, laying the intellectual groundwork for CSR as a field of ethics and management (Carroll, 1999).

By the 1980s, CSR had evolved further into a strategic business imperative. Academic voices like Peter Drucker argued that businesses could convert social problems into economic opportunities, rather than treating CSR as mere charity (Chakraborty, 2015). Simultaneously, the rise of Stakeholder Theory emphasized that corporations have obligations not only to shareholders but also to employees, communities, customers, and the environment (Jamali, 2008).

A major turning point came in 1987 with the publication of the Brundtland Report, as this report popularized the concept of sustainable development. It strongly influenced how CSR was understood in relation to environmental and social sustainability. With sustainability being first defined and including economic, sociocultural and environmental responsibilities, the world was given a framework for how to approach sustainability initiatives and practices.

Into the 1990s and 2000s, CSR gained institutional structure through additional global frameworks. The **Global Reporting Initiative (GRI)** was founded in 1997 by Ceres and the Tellus Institute, with support from UNEP. It published its first full sustainability reporting guidelines in 2000, helping companies systematically disclose social, environmental, and governance impacts (Global Reporting Initiative, 2021). Meanwhile, in 2000 the **UN Global Compact** was launched, calling on companies worldwide to adopt principles in human rights, labor, environment, and anti-corruption (UN Global Compact, 2023). Later, **ISO 26000** was published in 2010, offering guidance on social responsibility, developed through a multi-stakeholder process that included industry, government, non-governmental organizations, labor, consumers, and academics (ISO, 2010; UN Global Compact, 2007).

These frameworks underscore how CSR has transformed throughout the years from a supplemental philanthropic activity into a structured, strategic, and globally coordinated effort embedded within corporate governance, ethics, and long-term planning.

## 12.3 Drivers of CSR in the Modern Economy

The growing emphasis on CSR in the modern economy is shaped by a broader shift toward sustainability-driven governance, where businesses are increasingly evaluated not just on financial performance but also on their social and environmental impact. Global challenges such as climate change, biodiversity loss, supply-chain inequality, and resource scarcity have revealed that economic systems are deeply interdependent with ecological and social systems. As a result, companies face rising expectations to proactively address these issues rather than merely responding to crises. This transition has been accelerated by the digital age, which enables instantaneous information sharing and exposes unethical corporate behavior more quickly and publicly than ever before. In this environment, CSR becomes not only a moral imperative but a strategic approach to long-term business resilience, risk reduction, and competitiveness.

A particularly powerful driver of CSR is the exposure of poor labor conditions in global supply chains. High-profile tragedies, such as the collapse of the Rana Plaza factory in Bangladesh in 2013, where more than 1100 garment workers died, brought global attention to unsafe working environments, wage exploitation, and structural neglect (Business & Human Rights Resource Centre, 2014). The disaster underscored that corporate codes of conduct and conventional social audits were not sufficient to ensure worker safety. In response, many companies strengthened human-rights due diligence, adopted ethical sourcing standards, and committed to more robust supply chain monitoring. These labor abuses create reputational, legal, and operational risks, prompting firms to integrate labor rights into their CSR strategies as a central concern.

Several factors explain the growing importance of CSR:

- *Consumer Expectations*
  Conscious consumerism is a major driver of CSR. Multiple global surveys find a large share of consumers factor sustainability into purchasing choices. For example, McKinsey and NielsenIQ surveys report high shares of consumers saying they care about sustainability and that they will pay more for sustainably packaged products, and PwC's 2024 Voice of the Consumer survey found many consumers willing to pay a modest premium for sustainably produced goods (McKinsey & Company, 2023; NielsenIQ, 2023; PwC, 2024).
- *Employee Engagement*
  CSR is also vital for attracting and retaining talent. A survey conducted by the UN Global Compact found that younger workers, particularly Millennials and Generation Z, prefer employers who demonstrate ethical and sustainable practices (UN Global Compact, 2022).
- *Investor Pressure*
  Institutional investors are paying closer attention to Environmental, Social, and Governance (ESG) indicators, which often overlap with CSR metrics. The World Bank highlights that private sector investment aligned with ESG principles is a crucial driver of financing sustainable development (World Bank, 2020b).

- *Regulatory Scrutiny*
  Governments and international organizations are pushing for greater corporate accountability. Mandatory sustainability reporting is becoming more common, with countries such as the UK, EU members, and India requiring companies to disclose CSR or ESG-related activities.

Beyond these drivers, global regulatory developments related to labor rights are increasingly shaping CSR. For example, the European Union's Corporate Sustainability Due Diligence Directive (CSDDD), adopted in 2024, mandates companies to identify, prevent, or mitigate adverse impacts on human rights and the environment throughout their value chains, specifically including labor rights such as forced labor, child labor, and unsafe working conditions (EU Parliament & Council, 2023). Under this directive, companies must engage in due diligence and are expected to implement remediation measures where harm is identified (Business & Human Rights Resource Centre, 2024; Council of the EU, 2023). This growing legislative pressure demonstrates that CSR is no longer purely voluntary but increasingly tied to legal obligations, especially when labor exploitation is at stake.

CSR has increasingly expanded into Environmental, Social, and Governance (ESG) frameworks in the United States, as investors and firms seek to translate CSR principles into measurable criteria for evaluating corporate performance alongside financial returns. Supporters argue that ESG improves long-term risk evaluation by identifying exposures such as climate-related financial risk, labor practices, and regulatory instability, and it has been widely adopted by institutional investors and asset managers as a practical way to operationalize CSR in financial markets. However, ESG has also become highly politicized, with critics in some U.S. states arguing that it introduces ideological considerations into financial decision making and may conflict with the responsibility to prioritize financial returns for investors or not utilize funds effectively. This has led to legislative efforts in several states to restrict or penalize the use of ESG criteria in public investment decisions, creating a fragmented regulatory landscape that varies sharply across jurisdictions. As a result, ESG in the United States has become less a unified extension of CSR and more a contested political issue, exposing deeper divisions over whether corporate responsibility should prioritize shareholder returns alone or also explicitly incorporate environmental and social outcomes.

## 12.4 CSR and the Sustainable Development Goals (SDGs)

The United Nations SDGs provide a global framework for CSR. Many companies align their strategies with the 17 SDGs, recognizing that business success is tied to societal progress.

Examples include:

- *SDG 8 (Decent Work and Economic Growth):* Many companies adopt initiatives around fair wages, workplace safety, and inclusive employment.

- *SDG 12 (Responsible Consumption and Production):* Firms reduce waste, improve resource efficiency, and rethink product lifecycles.
- *SDG 13 (Climate Action):* Corporations commit to net-zero emissions, invest in renewable energy, or remove carbon from the atmosphere.

The United Nations Global Compact advances corporate responsibility through its Ten Principles on human rights, labor, environment, and anti-corruption. As a voluntary leadership platform, it brings together tens of thousands of participants worldwide. In recent reporting the initiative describes more than 20,000 participant companies and other organizations in over 160 countries, signaling its continued growth as a global convening platform for corporate sustainability. The Compact's Ten Principles remain a practical entry point for companies seeking to align business operations with international norms (UN Global Compact, 2025).

## 12.5 Highlights of CSR in Practice

*Unilever and Sustainable Living Plan* Unilever's CSR strategy integrates sustainability into every part of its supply chain. The company has pledged to reduce its environmental footprint by half while improving health and well-being for over a billion people. It aligns directly with SDGs such as poverty reduction (SDG 1), clean water (SDG 6), and climate action (SDG 13).

*Microsoft's Carbon Negative Pledge* In 2020, Microsoft announced its commitment to become carbon negative by 2030 and to remove all historical carbon emissions by 2050. This demonstrates how corporations can set ambitious climate goals in alignment with SDG 13 and contribute to global sustainability (Microsoft, 2020; United Nations, 2023).

*CSR in India* India became one of the first countries to mandate CSR through the Companies Act of 2013, which requires large firms to spend at least 2% of their profits on CSR initiatives. Projects include rural education, healthcare programs, and renewable energy investments (World Bank, 2020a; Companies Act, 2013; India Briefing 2014).

## 12.6 The Role of Reporting and Certification in CSR

Transparency is critical for CSR credibility, and reporting frameworks help ensure that companies' CSR efforts are genuine, measurable, and verifiable.

- *Global Reporting Initiative (GRI):* The GRI standards provide widely adopted guidelines for sustainability reporting, allowing companies to disclose their impacts on people, planet, and economic performance. They are the most widely used sustainability reporting standards. Since 2020 more than 10,000 reporting organizations in over 100 countries have used or referenced the GRI Standards to prepare their sustanability disclosures, making GRI the most widely adopted global reporting framework (GRI, 2022).

- *UN Global Compact: Communication on Progress (CoP):* As part of its membership commitments, companies must submit an annual CoP report, detailing how they implement the Ten Principles and integrate CSR into their strategy (UN Global Compact, 2022). This process holds companies publicly accountable, reducing the risk of greenwashing or "CSR-washing."

These mechanisms reduce the risk of "greenwashing" and ensure that CSR claims are evidence-based. This assists with creating consumer trust and encouraging more of the public to engage with companies who follow sustainable practices in their business.

## 12.7 CSR and Consumer Education

CSR and consumer education are increasingly recognized as complementary mechanisms by which firms contribute to public welfare while strengthening market relationships. CSR extends beyond regulatory compliance to proactive efforts that inform, protect, and empower consumers. This is done through transparent labeling, health and safety campaigns, product redesign, or direct educational programs. These efforts matter because consumer knowledge strongly shapes health behaviors.

For example, countries that do not have educational programs associated with informing the public of the risks of smoking have higher rates of smoking, which shows the connection with education and access to information to public behavior (OECD, 2023). Large public-health studies consistently show that people with higher levels of education are much less likely to smoke than those with less education (Tomioka et al., 2020). In the United States, for example, surveillance data show tens of millions of adults reporting current tobacco use and identify lower education as a risk correlate for continued tobacco use (CDC, 2024). These patterns signal that raising public knowledge about risks, cessation resources, and how to interpret product information can reduce harmful behaviors. Firms that incorporate consumer education into CSR amplify those public-health gains, especially where government outreach is limited or slow to reach specific demographic groups.

Educational programs have also proven effective in changing other behaviors beyond smoking, demonstrating the broader potential of consumer education. For instance, public campaigns promoting seat belt use, such as "Click It or Ticket," have increased proper seat belt usage and reduced fatal injuries by about 50% (CDC, 2025a, 2025b). Similarly, initiatives targeting drunk driving have led to measurable improvements; in one study, self-reported rates of driving after drinking fell from approximately 42% to 31% following educational interventions (Cordis, n.d.). Programs that teach households and communities about energy conservation have also driven tangible results. For example, school-based energy education initiatives in Europe produced energy savings of 15–20% in some schools, while other programs achieved reductions of up to 25% through student-led monitoring and action (Cordis, n.d.). Nutrition education programs in schools and communities have helped students increase their daily intake of fruits and vegetables and reduce consumption of sugary drinks, contributing to healthier dietary habits (Nguyen et al., 2023). These examples illustrate that when knowledge is paired with practical guidance, people are more likely to adopt safer and more responsible behaviors, reinforcing the value of incorporating consumer education into CSR strategies.

Businesses are uniquely positioned to deliver timely, targeted consumer education at scale, and many have treated education as part of strategic CSR rather than a peripheral philanthropy. Global companies have run behavior-change programs grounded in evidence and partnerships. For example, Unilever's Lifebuoy brand has invested in school and community-based handwashing education and behavior change interventions, reaching millions and using partnerships with UNICEF and local actors to maximize impact (Unilever, 2020).

Academic research supports that CSR programs which educate consumers can build trust, perceived competence, and long-term loyalty. These are outcomes that benefit both public welfare and business performance. Importantly, corporate educational efforts are not a substitute for regulation, but should work with and beyond standard policy to continue societal improvements. When firms voluntarily provide clear nutrition or product-safety information, fund prevention campaigns, or design products that reduce harm, they can accelerate behavior change that regulation alone may not achieve quickly. Overall, the connection of CSR and consumer education shows how businesses can make an impact beyond the products and services they sell.

## 12.8 Challenges of CSR Implementation

CSR continues to face several important challenges despite measurable progress in adoption and visibility. One major barrier concerns inequitable access to CSR resources. Small and medium sized enterprises, especially those operating in emerging markets, often lack the financial and managerial capacity to implement CSR initiatives at the scale seen in multinational corporations. Research shows these markets frequently struggle to meet global sustainability standards because they operate with thinner margins and limited technical capacity, which restricts their ability to adopt environmental management systems or invest in community development programs (International Labour Organization, 2023). This gap increases the risk that CSR becomes concentrated among large firms while smaller businesses are excluded from global sustainability frameworks.

Another challenge relates to the measurement of CSR impact. Many CSR outcomes, particularly those tied to social equity, community empowerment, or long term environmental restoration, are difficult to quantify with precision. Academic studies show that companies often rely on inconsistent indicators or self reported metrics, which makes it difficult to compare outcomes across industries or regions.

Even among companies that follow sustainability standards, reporting can vary widely in specificity and measurable indicators, which limits the usefulness of CSR disclosures for investors and regulators (Global Reporting Initiative, 2023). The reality is that environmental improvements such as biodiversity recovery or social improvements such as educational gains often unfold slowly, which complicates short term reporting cycles.

A further complication is short termism within public companies. As firms listed on stock exchanges are often pressured to demonstrate strong quarterly financial performance, they may hesitate to invest in CSR initiatives that require several years before yielding measurable return. This short term performance pressure can deter firms from investing in projects such as renewable energy transitions, worker retraining programs, or long horizon infrastructure upgrades, even when these investments generate long term value and reduce risk (McKinsey & Company, 2023). This mismatch between short term expectations and long term sustainability needs continues to slow the pace of CSR adoption.

Regulatory ambiguity also presents a significant challenge. In India, for example, the Companies Act of 2013 introduced a requirement for qualifying companies to spend 2% of average net profits on CSR activities. However, government evaluations show that companies often misunderstand what qualifies as legitimate CSR expenditure, how to calculate eligible profits, or how to justify unspent CSR funds. The Ministry of Corporate Affairs reported that inconsistent interpretations of the law have led to compliance gaps, uneven reporting quality, and substantial funds remaining unspent, particularly among companies with limited administrative capacity (Ministry of Corporate Affairs, 2022). These issues illustrate that strong regulation alone cannot guarantee effective CSR unless clear guidance, administrative support, and enforcement mechanisms are also in place.

Taken together, these challenges demonstrate that advancing CSR requires not only corporate commitment but also improved regulatory clarity, better measurement systems, and expanded support for smaller firms that lack the capacity to participate fully in global sustainability frameworks.

## 12.9 Future Directions in CSR

Looking ahead, CSR is expected to evolve into a more innovation driven and measurable component of business strategy. One important development is the rapid growth of regenerative agriculture across global supply chains. Major companies such as PepsiCo and Nestlé have committed to supporting regenerative farming practices on millions of hectares of farmland by 2030, which reflects a significant shift toward soil restoration, biodiversity enhancement, and long term carbon sequestration (FAIRR, 2024). These changes indicate that sustainability is no longer treated as a peripheral effort but rather an integral part of operational planning and resource security.

At the same time, businesses are accelerating experimentation with circular product design. In the textile and apparel industry, for example, some companies are adopting modular construction techniques or using materials that can be fully recycled at the end of a product's lifespan. These approaches challenge the traditional linear production system and enable the reuse of valuable resources while also reducing waste (ClimateTech Digital, 2024). Parallel to these efforts, digital innovations such as product passports are gaining prominence. Digital product passports allow consumers, auditors, and regulators to trace materials and environmental impacts across an item's entire lifecycle, which improves transparency and strengthens the credibility of sustainability reporting (Illán García et al., 2024).

CSR is also expanding through new types of social innovation that link sustainability with employee wellbeing, community development, and fairer supply chain practices. Agricultural companies such as Cargill demonstrate this through programs like RegenConnect, which supports farmers with financial incentives, training, and tools for implementing climate beneficial practices including cover cropping and reduced tillage (World Grain, 2024). These programs show that forward looking CSR models increasingly combine environmental stewardship with community resilience and long term economic value.

As investors and consumers continue to demand stronger accountability, companies are beginning to tie financial incentives to verifiable sustainability outcomes. These developments suggest that the future of CSR will focus on creating business models that are regenerative, circular, inclusive, and transparent. Rather than merely minimizing harm, CSR is moving toward an approach where environmental improvement, social benefit, and economic performance reinforce one another and create durable value for both companies and society.

## 12.10 Wrap-Up

CSR is no longer a peripheral activity, as it has become central to how businesses operate in a globalized and interconnected world. Driven by consumer demand, investor scrutiny, and international frameworks, CSR provides a mechanism for companies to align profit-making with societal and environmental objectives. Through tools such as the Global Reporting Initiative, partnerships like the UN Global Compact, and commitments to the SDGs, corporations are reshaping their roles as stewards of both society and the environment. Modern CSR increasingly emphasizes innovation, transparency, and measurable impact. For instance, Microsoft's carbon-negative commitment and Unilever's handwashing education programs demonstrate how companies are integrating CSR into core operations while producing tangible social and environmental benefits. Similarly, PepsiCo and Nestlé's adoption of regenerative agriculture reflects a shift toward long-term resource stewardship that benefits both communities and ecosystems.

Despite these advances, challenges remain in ensuring authenticity, inclusivity, and long-term commitment. Small and medium-sized enterprises often face barriers to participation, while measuring social and environmental impact remains complex. Regulatory pressures are increasing, yet ambiguity persists in some jurisdictions, highlighting the need for clearer guidance and capacity building. Looking forward, CSR will likely continue to evolve through digital innovations such as product passports, circular economy models, and incentive structures that tie financial performance to verifiable social and environmental outcomes. When fully embedded into corporate strategy, CSR does more than enhance reputation, as it drives sustainable innovation, strengthens stakeholder trust, and contributes meaningfully to global efforts to achieve environmental sustainability, social equity, and ethical governance. Ultimately, CSR represents both a strategic necessity and a moral commitment, enabling businesses to create durable value for society, the planet, and the organizations themselves.

## References

Association of Corporate Citizenship Professionals. (2023). *Corporate social responsibility: A brief history*. https://accp.org/resources/csr-resources/accp-insights-blog/corporate-social-responsibility-brief-history/

Business & Human Rights Resource Centre. (2014). *Rana Plaza building collapse*. Retrieved April 2013, from https://www.business-humanrights.org/en/latest-news/rana-plaza-building-collapse-april-2013/

Business & Human Rights Resource Centre. (2024). *EU corporate sustainability due diligence directive summary*. https://www.business-humanrights.org/documents/39897/EU_CSDDD_summary_3.17.24.pdf

Carroll, A. B. (1999). Corporate social responsibility: Evolution of a definitional construct. *Business & Society, 38*(3), 268–295.

CDC. (2024). *Current cigarette smoking among adults in the United States*. Centers for Disease Control and Prevention. https://www.cdc.gov/tobacco/php/data-statistics/adult-data-cigarettes/index.html

CDC. (2025a, January 31). *Facts about seat belt use*. Centers for Disease Control and Prevention. https://www.cdc.gov/seat-belts/facts/index.html CDC

CDC. (2025b, July 28). *MVPICCS intervention: Seat belt enforcement campaign*. Centers for Disease Control and Prevention. https://www.cdc.gov/transportation-safety/calculator/highvis.html

Chakraborty, U. K. (2015). Developments in the concept of corporate social responsibility (CSR). *The Researchers' International Research Journal, 1*(1), 23–45.

ClimateTech Digital. (2024). *Top 10 circular design strategies*. https://climatetechdigital.com/top-10/top-10-circular-design-strategies

Companies Act. (2013). *Companies Act No. 18 of 2013*. Government of India. https://www.indiacode.nic.in/handle/123456789/2114

Cordis (n.d.). *Supporting European schools towards sustainability awareness and energy efficiency (GAIA project)*. European Commission. https://cordis.europa.eu/article/id/411418-supporting-european-schools-towards-sustainability-awareness-and-energy-efficiency

European Council & Parliament. (2023, December 14). *Corporate sustainability due diligence: Council and parliament strike deal to protect environment and human rights*. Press release. https://www.consilium.eu.europa.eu/en/press/press-releases/2023/12/14/corporate-sustainability-due-diligence-council-and-parliament-strike-deal-to-protect-environment-and-human-rights/

FAIRR. (2024, June). *Food sector making more promises than progress on regenerative agriculture.* https://www.fairr.org/news-events/press-releases/food-sector-making-more-promises-than-progress-on-regenerative-agriculture

GRI. (2021). *Consolidated set of GRI standards.* Global Reporting Initiative. https://www.globalreporting.org/standards/

GRI. (2022). *GRI history and consolidated set of standards.* Global Reporting Initiative. https://www.globalreporting.org/standards/

GRI. (2023). *Consolidated GRI standards 2021 update.* Global Reporting Initiative. https://www.globalreporting.org/standards

Illán García, I., Muñoz Escoí, F. D., & Arjona Aroca, J. (2024). *Digital product passport management with decentralised identifiers and verifiable credentials. arXiv.* https://arxiv.org/abs/2410.15758

ILO. (2023). *Empowering small businesses: Strengthening sustainability in small and medium sized enterprises.* International Labour Organization. https://www.ilo.org

India Briefing. (2014, March 12). *India's corporate social responsibility mandate:* The Companies Act 2013. https://www.india-briefing.com/news/indias-corporate-social-responsibility-mandate-companies-act-2013-8057.html/

ISO. (2010). *ISO 26000:2010 Guidance on social responsibility.* International Organization for Standardization. https://www.iso.org/standard/42546.html

Jamali, D. (2008). A stakeholder approach to corporate social responsibility: A fresh perspective into theory and practice. *Journal of Business Ethics, 82*(1), 213–231.

McKinsey & Company. (2023). *The economic forces shaping sustainable investing.* https://www.mckinsey.com

Microsoft. (2020). *Microsoft announces it will be carbon negative by 2030.* https://news.microsoft.com/source/2020/01/16/microsoft-announces-it-will-be-carbon-negative-by-2030/

Ministry of Corporate Affairs. (2022). *Report on corporate social responsibility in India.* Government of India. https://www.mca.gov.in

Nguyen, T., Hasan, M., & Khan, M. (2023). Efficacy of a school-based education intervention on the consumption of fruits, vegetables and carbonated soft drinks among adolescents. *Public Health Nutrition, 26*(12), 3112–3121. https://pubmed.ncbi.nlm.nih.gov/37781771

NielsenIQ. (2023). *Consumer sustainability trends.* https://nielseniq.com/global/en/insights/infographic/2023/consumer-sustainability-trends-40-claims-driving-sustainable-consumers-to-buy/

OECD. (2023). *Health at a glance 2023: Smoking.* Organisation for Economic Co-operation and Development. https://www.oecd.org/health/health-at-a-glance/

PwC. (2024). *Voice of the consumer* 2024. https://www.pwc.com/gx/en/news-room/press-releases/2024/pwc-2024-voice-of-consumer-survey.html

Satell Institute. (2023). *A history of corporate social responsibility: Concepts and principles.* https://www.satellinstitute.org/whitepapers/a-history-of-corporate-social-responsibility-concepts-and-practices

Tomioka, K., Kurmantani, N. & Saeki, K. (2020). *The association between education and smoking: A systematic review and meta-analysis.* BMC Public Health. https://www.ncbi.nlm.nih.gov/pmc/articles/PMC7025916/

UN Global Compact. (2022). *Annual report 2022: Delivering impact through principles and partnerships.* United Nations. https://www.unglobalcompact.org/library/6191

UN Global Compact. (2023). *The ten principles of the UN global compact.* https://www.unglobalcompact.org/what-is-gc/mission/principles

UN Global Compact. (2025). *The world's largest corporate sustainability initiative.* https://unglobalcompact.org/what-is-gc.

Unilever. (2020). *Help Lifebuoy change the world's handwashing habits.* https://www.unilever.com/news/news-search/2020/help-lifebuoy-change-the-worlds-handwashing-habits-forever/

United Nations. (2023). *Microsoft: Carbon negative goal.* https://unfccc.int/climate-action/un-global-climate-action-awards/climate-neutral-now/microsoft-carbon-negative-goal.

United Nations Global Compact. (2007). *The UN Global Compact: Operational guide for medium-sized companies*. United Nations Global Compact Office. https://unglobalcompact.org/library/226

World Bank. (2020a). *Transforming business for a sustainable future: The role of the private sector in achieving the SDGs*. World Bank Group. https://documents.worldbank.org/en/publication/documents-reports/documentdetail/638991591339530445/transforming-business-for-a-sustainable-future-the-role-of-the-private-sector-in-achieving-the-sdgs

World Bank. (2020b). *Climate finance and ESG integration in emerging markets*. World Bank. https://www.worldbank.org/en/topic/climatechange

World Grain. (2024). *Agribusiness cultivating value through sustainability: Cargill's RegenConnect initiative*. https://www.world-grain.com/articles/21985-agribusiness-cultivating-value-through-sustainability

# Chapter 13
# Measuring Progress

## 13.1 How Much Progress Has Been Made?

Measuring progress in sustainability is crucial for ensuring that efforts to protect the environment, enhance social well-being, and foster responsible economic growth are actually producing results. Without clear benchmarks, reliable data, and consistent evaluation, sustainability risks becoming a vague aspiration rather than a measurable reality. Governments, businesses, and communities all need mechanisms to monitor their sustainability initiatives, assess their impacts, and identify areas for improvement.

At the corporate level, this is often done through **Key Performance Indicators (KPIs)**, benchmarks, and frameworks such as **SMART goals** (Specific, Measurable, Achievable, Relevant, Time-bound). These tools provide tangible metrics that make sustainability measurable rather than theoretical. Not only is this vital for accountability and transparency, but also so the organization can learn and continue to improve (World Bank, n.d.). Similarly, international organizations such as the United Nations emphasize data-driven approaches for monitoring progress, particularly through the Sustainable Development Goals (SDGs), which include over 230 unique indicators used to track global progress (United Nations, 2022; United Nations Statistics Division, 2025).

**Benchmarking** involves comparing an organization's sustainability performance against internal goals, industry standards, or global best practices. It allows businesses to evaluate where they stand relative to peers, competitors, and global norms.

For instance, the United Nations Environment Programme (UNEP) encourages businesses to align their benchmarks with the SDGs and national climate commitments, such as those outlined in the Paris Agreement (UNEP, 2022). Benchmarking also helps in identifying leaders and laggards in sustainability, creating both accountability and opportunities for knowledge sharing.



N. Hugo, *Towards Sustainability*, https://doi.org/10.1007/978-3-032-29339-8_13

An example of benchmarking can be seen with renewable energy in the EU. It established binding benchmarks for renewable energy use, setting targets in 2019 for member states to reach 32% renewable energy consumption by 2030 (European Commission, 2019).. This was then raised in 2023 to 42.5% by 2030 (European Commission, 2023). Countries then report annually on their progress toward these benchmarks, providing a model for how large-scale sustainability performance can be systematically measured.

## 13.2 Key Performance Indicators (KPIs) for Sustainability

**KPIs** are one of the most common tools for measuring sustainability progress. Hey provide quantifiable metrics across environmental, social, and economic dimensions, enabling organizations to evaluate performance, identify areas for improvement, and demonstrate accountability to stakeholders. By establishing clear KPIs, businesses and institutions can move beyond general sustainability statements to measurable, evidence-based action.

Common sustainability KPIs include:

- *Environmental:* Greenhouse gas (GHG) emissions, energy efficiency, water usage, waste recycling rates, and biodiversity impacts. These indicators measure the direct ecological footprint of organizational operations and supply chains.
- *Social:* Employee satisfaction, workforce diversity, community investment, and health and safety records. These metrics reflect an organization's social responsibility, internal culture, and engagement with surrounding communities.
- *Economic:* Revenue generated, cost savings, investments and ripple effects to funded opportunities, financial resilience through long-term sustainable strategies, and job creation.

An organization that assists with educating and providing resources about KPIs is the Global Reporting Initiative (GRI). The GRI provides standardized KPIs for organizations to report their sustainability impacts across these categories, creating consistency and comparability across industries (GRI, 2021). By adhering to these frameworks, organizations can ensure that data on environmental, social, and economic performance is credible, transparent, and comparable over time.

KPIs also serve as the foundation for strategic planning and continuous improvement. In addition to measuring current performance, they allow organizations to set targets, track progress over multiple reporting periods, and identify correlations between sustainability initiatives and overall organizational performance. For example, tracking the financial returns of renewable energy investments or energy efficiency upgrades helps quantify the economic benefits of environmental initiatives, providing evidence that sustainability can enhance competitiveness and profitability. Similarly, monitoring workforce diversity alongside productivity and retention rates can demonstrate how social responsibility initiatives contribute to operational resilience.

KPIs are not only relevant to corporations but also to governmental and nonprofit organizations.

National governments may track economic KPIs such as green job creation, sustainable investment flows, or GDP contributions from low-carbon industries. Nonprofits can assess the financial efficiency of programs while measuring social outcomes, ensuring that resources are deployed effectively to achieve sustainability goals.

The company Unilever uses sustainability KPIs to track its "Sustainable Living Plan," which monitors everything from greenhouse gas reductions to improved livelihoods in supply chains. By 2020, the company reported cutting $CO_2$ emissions from energy in production by 65% compared to 2008 levels, while achieving significant progress in social metrics such as supplier labor standards (Unilever, 2020). This example illustrates how KPIs can integrate environmental, social, and economic dimensions, providing a comprehensive view of an organization's contribution to sustainable development.

KPIs are also critical for stakeholder engagement. Investors increasingly assess economic KPIs to evaluate both risk management and long-term value creation. Customers and employees similarly rely on KPI reporting to determine whether organizations align with their ethical expectations. Over time, robust KPIs help shift sustainability from a peripheral consideration into a core element of business strategy, driving evidence-based decisions and creating measurable impact across environmental, social, and economic dimensions.

## 13.3 SMART Goals and Sustainability Measurement

The **SMART Goals framework** stives to makes sure goals are Specific, Measurable, Achievable, Relevant, and Time-bound. This ensures that goals are practical and trackable rather than aspirational.

For example:

- Instead of stating, *"We will reduce emissions,"* a SMART goal would be, *"Reduce Scope 1 and Scope 2 carbon emissions by 25% by 2030, compared to 2015 levels."*
- Instead of claiming, *"We will improve water efficiency,"* a SMART goal could be, *"Reduce freshwater withdrawals by 15% per unit of production by 2025."*

The World Bank highlights that SMART goals in sustainability allow policymakers and businesses to align investments with measurable development outcomes, making them critical for financing and accountability (World Bank, 2021).

## 13.4 Tools and Methodologies for Measuring Sustainability

Measuring sustainability progress requires a combination of quantitative and qualitative tools that assess environmental, social, and economic impacts. These methodologies help organizations, governments, and communities translate abstract sustainability goals into actionable metrics. By employing robust tools, stakeholders can identify strengths, weaknesses, and opportunities, track progress over time, and support evidence-based decision-making.

- *Carbon Footprint Calculators*
  Carbon accounting tools quantify greenhouse gas (GHG) emissions generated by products, operations, or entire organizations. These calculators allow businesses and governments to identify key sources of emissions, evaluate reduction strategies, and set measurable targets for mitigation. For example, carbon footprint methodologies follow internationally recognized protocols, such as the Greenhouse Gas Protocol or guidance provided by the UNFCCC (United Nations Framework Convention on Climate Change), which ensures consistency in reporting across countries and industries (UNFCCC, 2018). Beyond corporate operations, carbon calculators are increasingly applied at the product level, supply-chain level, and even for individual consumption patterns, enabling precise measurement of emissions intensity per unit of output or per capita.
- *Social Impact Measurement*
  Assessing social outcomes is critical for a holistic understanding of sustainability. Tools such as Social Return on Investment (SROI), social audits, or community engagement surveys evaluate improvements in equity, inclusion, health, education, and local well-being (OECD, 2019). Social impact measurement allows organizations to quantify the effectiveness of programs, compare interventions, and communicate results to stakeholders transparently. Metrics may include job creation in underserved communities, improvements in access to education or healthcare, workforce diversity, or the economic effects of community development initiatives
- *Economic Sustainability Metrics*
  Economic measurement focuses on financial performance linked to sustainability objectives, capturing how environmental and social actions affect economic outcomes. Key metrics may include revenue generated from sustainable products or services, cost savings from energy efficiency or circular economy initiatives, and financial resilience through long-term sustainable investments. Economic KPIs can also reflect indirect impacts, such as local economic growth stimulated by green investments or sustainable job creation. For governments, economic sustainability metrics may encompass GDP contributions from low-carbon sectors, tax revenues from sustainable industries, or the financial viability of ecosystem service preservation. By linking sustainability to measurable economic performance, organizations demonstrate that responsible practices can align with growth, profitability, and resilience.

- *Composite Indices*
  Composite indices integrate multiple indicators to provide a broader picture of sustainability performance. Examples include the Human Development Index (HDI), which combines health, education, and income metrics, and the Environmental Performance Index (EPI), which aggregates environmental health and ecosystem vitality data (UNDP, 2020). These indices allow policymakers and organizations to assess systemic trends, identify priority areas for intervention, and compare performance across countries or regions.

## 13.5 Examples of Progress Tracking

Sustainability measurement tools and methodologies are applied at national and global levels, highlighting the value of KPIs, benchmarking, and composite indices. We went in detail about how the United Nations tracks the SDGs, but the following examples illustrate how other organizations track and measure progress.

At the national level, Rwanda has implemented a circular economy action plan with measurable KPIs that track waste reduction, recycling rates, and the establishment of eco-industrial parks. By 2022, Rwanda became a leader in Africa in banning single-use plastics and systematically measuring national waste management improvements (UNEP, 2022). Economic KPIs also played a role, including job creation in recycling and green industries and the economic contributions of sustainable infrastructure projects, demonstrating how environmental and social progress can align with economic development.

The World Bank's comprehensive wealth framework provides another example of integrated sustainability measurement. It tracks the total wealth of nations, including produced capital, natural capital, human capital, and net foreign assets (World Bank, 2018). By incorporating economic, social, and environmental components, this approach allows policymakers to assess whether growth is sustainable over the long term, rather than relying solely on short-term GDP figures. For instance, countries with declining natural capital may face long-term economic risks even if short-term production appears strong.

At the corporate level, the use of KPIs for sustainability enables companies to monitor operational impacts and demonstrate accountability to stakeholders. For example, sustainability KPIs can track reductions in greenhouse gas emissions, improvements in workforce diversity, and financial benefits from energy efficiency or sustainable supply-chain investments. By comparing performance against internal goals or industry benchmarks, companies can identify leaders and laggards, create incentives for improvement, and share best practices. Economic KPIs in particular, such as revenue from sustainable products, cost savings, and long-term investment returns, provide evidence that sustainability initiatives can enhance profitability while contributing to broader societal goals.

These examples demonstrate how measurement frameworks, when applied consistently, provide accountability, guide resource allocation, and support evidence-based decision-making. By integrating environmental, social, and economic KPIs, stakeholders can track progress comprehensively, ensure transparency, and translate sustainability commitments into tangible outcomes.

## 13.6 Challenges in Measuring Progress

Despite the availability of KPIs, benchmarking tools, and composite indices, measuring sustainability progress remains a complex task with multiple challenges that affect accuracy, comparability, and long-term reliability. These challenges span environmental, social, and economic dimensions, highlighting the need for careful design, implementation, and interpretation of sustainability metrics.

- *Data Gaps:* Reliable and consistent data are essential for accurate measurement, yet many regions and organizations face limitations. Developing countries often lack the infrastructure, technical capacity, or financial resources to collect comprehensive environmental, social, or economic data. Similarly, smaller organizations may struggle to gather the detailed internal information required to calculate meaningful KPIs, leading to incomplete or inconsistent reporting. Data gaps can result in underestimation of impacts or missed opportunities for intervention, reducing the effectiveness of sustainability initiatives.
- *Standardization Issues:* Another major challenge is the lack of uniformity in KPI definitions, measurement methods, and reporting frameworks. Organizations across different sectors may report similar indicators using different methodologies, making direct comparison difficult. For example, two companies reporting on "energy efficiency" may calculate it differently, or national governments may measure "green jobs" using divergent criteria. These inconsistencies can hinder benchmarking, reduce transparency, and limit the ability of stakeholders to accurately evaluate performance. Global initiatives, such as the Global Reporting Initiative and the SDG indicators, attempt to standardize definitions, but adoption is not universal, and variations in implementation persist.
- *Greenwashing Risks:* Without independent verification, organizations may manipulate data or selectively report metrics to appear more sustainable than they truly are. Greenwashing can occur across environmental, social, and economic dimensions, from overstating reductions in greenhouse gas emissions to exaggerating community investments or economic benefits of sustainability initiatives. This practice undermines stakeholder trust, reduces the credibility of reported metrics, and can distort decision-making. Transparency, third-party audits, and adherence to established reporting standards are essential to mitigate these risks.
- *Time Horizons:* Sustainability outcomes often require long-term observation, which complicates short-term measurement. Environmental recovery, biodiversity restoration, or economic resilience from sustainable investments may take years or decades to fully materialize. Short-term data may not capture

delayed impacts, leading to premature conclusions about the effectiveness of interventions. Similarly, social initiatives, such as workforce development programs or community engagement efforts, may produce gradual changes that are difficult to quantify in conventional reporting periods.
- *Integration of Multi-Dimensional Metrics:* Tracking progress across environmental, social, and economic dimensions requires integrating disparate types of data. While environmental indicators often rely on physical measurements, social metrics may involve surveys or qualitative assessment, and economic KPIs may focus on financial performance or investment returns. Combining these metrics into a coherent picture of progress presents methodological and analytical challenges, especially when data quality and availability vary across dimensions.
- *Contextual Differences:* Benchmarking progress across regions or sectors must account for contextual differences in regulatory environments, market structures, cultural norms, and resource availability. A KPI that is feasible and meaningful in one country or industry may not be applicable elsewhere. These contextual factors can complicate the interpretation of results and the comparison of performance between organizations or nations.

Overall, these challenges illustrate that sustainability measurement is not simply a technical exercise but requires careful attention to methodology, verification, and interpretation. Addressing these obstacles is critical for ensuring that sustainability metrics accurately reflect real progress and support effective decision-making across environmental, social, and economic dimensions.

## 13.7 Future Directions

Addressing the challenges of measuring sustainability progress requires coordinated efforts to improve data quality, consistency, and interpretability, as well as to enhance stakeholder confidence in reported results. One critical area of development is the continued standardization of metrics and reporting frameworks. International initiatives, such as the International Sustainability Standards Board and the Global Reporting Initiative, are working to unify definitions, methodologies, and disclosure requirements across industries and regions. Widespread adoption of these standards will help ensure that environmental, social, and economic indicators are comparable and transparent, enabling more effective benchmarking and evaluation of progress over time.

Another important direction is the integration of sustainability metrics with broader organizational decision-making. As businesses, governments, and nonprofit organizations increasingly recognize the strategic value of sustainability, progress measurement is becoming embedded within planning, budgeting, and risk management processes. Linking KPIs and benchmarking data to operational and financial performance allows organizations to better understand the economic, social, and environmental trade-offs of different initiatives, and to prioritize actions that produce the greatest overall impact.

Collaboration and knowledge sharing are also likely to expand as a key feature of future sustainability measurement. Multi-stakeholder partnerships between governments, international organizations, research institutions, and civil society can provide access to specialized expertise, shared data platforms, and verified indicators. Such collaborations support capacity building in regions or organizations with limited resources, while fostering a global perspective on sustainability progress. At the same time, greater emphasis on transparency and verification will help reduce greenwashing risks. Independent audits, peer reviews, and publicly accessible reporting will increase accountability, strengthening trust among investors, consumers, and communities.

Finally, there is a growing recognition of the need for adaptive and context-sensitive approaches to measurement. Sustainability outcomes often manifest over long time horizons and may be influenced by local conditions, economic structures, and cultural norms. Future measurement strategies will need to accommodate these differences, using flexible indicators, longitudinal tracking, and integrated assessments to capture progress accurately. By improving the rigor, accessibility, and usability of sustainability data, these approaches will empower decision-makers, enable evidence-based policy and investment choices, and ultimately help ensure that sustainability commitments translate into tangible, measurable improvements for the environment, society, and the economy.

## 13.8 Wrap-Up

Measuring progress is fundamental to translating sustainability commitments into tangible outcomes. Without clear benchmarks, standardized KPIs, and frameworks such as SMART goals, sustainability risks remaining an abstract concept rather than a measurable reality. Tools and methodologies including carbon footprint calculators, social impact assessments, economic metrics, and composite indices enable organizations, governments, and communities to track performance across environmental, social, and economic dimensions. Consistent application of these tools provides insight into strengths, weaknesses, and opportunities for improvement while supporting evidence-based decision-making and strategic resource allocation.

KPIs play a central role in sustainability measurement by allowing stakeholders to monitor operational impacts, assess progress relative to targets or industry standards, and communicate accountability. They provide a structured way to evaluate how initiatives contribute to environmental stewardship, social responsibility, and economic resilience. KPIs also support continuous improvement, enabling organizations to track trends over time, adjust strategies as needed, and identify correlations between sustainability efforts and broader organizational or societal outcomes.

Despite advances in measurement, challenges remain including data gaps, inconsistencies in KPI definitions, risks of selective reporting, long-term outcome measurement, integration of multi-dimensional metrics, and contextual variations across sectors and regions. Addressing these challenges is essential to ensure that KPI-based measurement accurately reflects real progress. Future approaches will require greater standardization, integration of KPIs into planning and decision-making, collaboration across stakeholders, and adaptive, context-sensitive methods that account for local conditions and long-term effects.

Ultimately, robust KPI-driven measurement transforms sustainability from a conceptual goal into actionable insight. By linking environmental, social, and economic dimensions, these measurement systems provide accountability, guide informed decision-making, and ensure that sustainability initiatives generate meaningful and measurable impacts for both present and future generations.

## References

European Commission. (2019). *Renewable energy directive (RED II): Binding target for renewable energy*. European Commission. https://energy.ec.europa.eu/topics/renewable-energy/renewable-energy-directive-targets-and-rules/renewable-energy-targets_en

European Commission. (2023, November 20). *Renewable Energy Directive (EU/2023/2413)*. https://energy.ec.europa.eu/topics/renewable-energy/renewable-energy-directive-targets-and-rules/renewable-energy-targets_en

GRI. (2021). *Consolidated set of GRI standards*. Global Reporting Initiative. https://www.globalreporting.org/standards/gri-standards-download-center/

OECD. (2019). *Social impact measurement for the social and solidarity economy*. Organisation for Economic Co-operation and Development. https://www.oecd.org/en/publications/social-impact-measurement-for-the-social-and-solidarity-economy_d20a57ac-en.html

UNDP. (2020). *Human development report 2020: The next frontier*. United Nations Development Programme. https://hdr.undp.org/system/files/documents/hdr2020.pdf

UNEP. (2022, January). *UNEP in 2021: Planetary action: Climate, nature, chemicals & pollution*. United Nations environment Programme. https://wedocs.unep.org/20.500.11822/37946 UN Environment Document Repository.

UNFCCC. (2018). *The Paris agreement*. United Nations Framework Convention on Climate Change. https://unfccc.int/process-and-meetings/the-paris-agreement

Unilever. (2020). *Unilever Sustainable Living Plan: Summary of 10 years' progress (2010–2020)*. Unilever PLC. https://www.unilever.com/files/4eb5788f-5ec4-43ad-88f3-ce139fa8d3a2/uslp-summary-of-10-years-progress.pdf

United Nations. (2022). *The sustainable development goals report 2022*. United Nations. https://www.un.org/sites/un2.un.org/files/the-sustainable-development-goals-report-2022.pdf

United Nations Statistics Division. (2025). *SDG indicators: Global indicator framework for the sustainable development goals*. United Nations. https://unstats.un.org/sdgs/indicators/indicators-list/.

World Bank. (2018). *The changing wealth of nations 2018: Building a sustainable future*. World Bank. https://openknowledge.worldbank.org/entities/publication/ef8b5373-1efd-5aec-b0db-ca4d7e9fd321

World Bank. (n.d.). *Measuring and reporting results in the World Bank*. https://thedocs.worldbank.org/en/doc/7b776fece25d919127d83a8a6a396da2-0290032021/original/Measuring-and-reporting-results-factsheet.pdf

World Bank Group. (2021). *Climate Change Action Plan 2021–2025: Supporting green, resilient, and inclusive development*. World Bank Group. https://openknowledge.worldbank.org/entities/publication/667f5ac5-6b4a-5d37-bc66-6ad8c8837f57

# Appendices

## 1.1 Appendix A. Glossary of Terms

**Attitudes:** Individual beliefs and evaluations about a behavior, including whether it is seen as positive, beneficial, or worthwhile. Positive attitudes toward a sustainable action increase the likelihood of performing that behavior.

**Benchmarking:** A method of comparing an organization's performance against internal objectives, industry standards, or best practices. Benchmarking identifies gaps, highlights leaders and laggards, and facilitates knowledge sharing for continuous improvement.

**Brundtland Report:** A 1987 United Nations report led by Gro Harlem Brundtland that introduced the widely used definition of sustainable development, which was "meeting the needs of the present without compromising the ability of future generations to meet their own needs". It helped bring global attention to sustainability and sustainable development, as well as broaden the scope of sustainability initiatives beyond the environment to include economic and socio-cultural issues.

**Circular Economy:** An economic model that prioritizes resource efficiency, waste reduction, and the continuous reuse or regeneration of materials. Unlike linear "take-make-dispose" systems, circular economies aim to keep products and resources in use, regenerate natural systems, and create sustainable economic opportunities.

**Climate change:** Climate change refers to long-term shifts in temperature, precipitation, and weather patterns caused mainly by human activities such as the burning of fossil fuels. These changes include more frequent storms, flooding, droughts, and other extreme weather events. Climate change affects all three pillars of sustainability because it disrupts ecosystems, limits resources, threatens economic stability, and forces social and cultural adjustments.



N. Hugo, *Towards Sustainability*, https://doi.org/10.1007/978-3-032-29339-8

**Corporate Social Responsibility (CSR):** A strategic business approach in which companies integrate social, environmental, and ethical considerations into their operations beyond profit-making. CSR encompasses initiatives to protect human rights, promote equitable labor practices, reduce environmental harm, and ensure ethical governance.

**Cradle-to-Cradle:** A design and production approach that treats products as part of a regenerative system. Unlike traditional "cradle-to-grave" models, Cradle-to-Cradle ensures that products can be fully reused, recycled, or biodegraded, minimizing waste and supporting circular economy principles.

**Cultural Heritage:** This refers to the traditions, languages, customs, sacred places, and shared knowledge that connect people to their history and identity. Cultural heritage creates belonging and strengthens community resilience. In this chapter, it includes both physical places like historic sites and intangible practices like storytelling, cuisine, and rituals. Cultural heritage must be protected because it preserves identity and supports sustainable social development.

**Developed Nation:** A country with a high level of economic prosperity, advanced infrastructure, strong institutions, and well-established social systems. Developed nations have already undergone extensive industrial growth, allowing them to adopt sustainable technologies and climate adaptation strategies more easily.

**Developing Nation:** A country that is still building its economic systems, infrastructure, and social services while working toward higher levels of stability and prosperity. Developing nations often face constraints such as limited financial resources, restricted access to technology, and greater vulnerability to environmental and economic challenges.

**Efficiency:** The ability to use resources in a way that maximizes output while minimizing waste, cost, and environmental impact. Efficiency focuses on getting the greatest possible value from limited inputs such as energy, materials, time, and labor. It supports long-term prosperity by conserving resources, reducing expenses, and helping societies manage scarcity.

**Equity:** Fairness in the distribution of economic opportunities, resources, and benefits across all members of society. Equity promotes inclusion by ensuring that people have access to education, finance, jobs, and essential services. It seeks to reduce inequality, expand opportunities for disadvantaged groups, and strengthen social and economic resilience.

**Global Reporting Initiative (GRI):** An international framework that provides guidelines for sustainability reporting, enabling companies to disclose their environmental, social, and governance impacts in a standardized and verifiable way, promoting transparency and accountability in CSR practices.

**Global warming:** Global warming refers specifically to the rise in the average surface temperature of the Earth. It is caused by the buildup of greenhouse gases in the atmosphere. Global warming is one part of climate change, with the warming acting as the trigger for many other climate-related impacts such as stronger storms and more intense droughts.

**Globalization:** The process through which countries become increasingly connected by exchanging goods, services, ideas, technology, capital, and cultural practices. Globalization shapes economic relationships between nations and influences how benefits and burdens are distributed across global supply chains.

**Green Deal:** A comprehensive policy framework created to guide economic transformation toward long-term sustainability. It connects environmental goals with social and economic progress by promoting clean energy, reducing emissions, supporting innovation, and improving quality of life. The Green Deal encourages cooperation across industries by linking climate action with fairness, worker protection, and regional development. In this chapter, it represents a political commitment to reshape entire systems so that future growth is both environmentally responsible and socially inclusive.

**Green Claims Directive:** A proposed European Union regulation that would require companies to provide scientific evidence for environmental claims made in marketing, ensuring that labels and claims are accurate, comparable, and verifiable across the EU. The directive aims to reduce greenwashing by enforcing transparency and accountability in sustainability-related marketing.

**Greenhouse gases:** Greenhouse gases are gases in the atmosphere that trap heat and make the planet warmer. Major greenhouse gases include carbon dioxide, methane, nitrous oxide, and fluorinated gases. Human activities that involve burning fossil fuels release large amounts of these gases. Reducing greenhouse gas emissions is essential for slowing global warming and addressing climate change.

**Greenwashing:** The practice of making misleading, exaggerated, or unsubstantiated claims about the environmental or social benefits of a product, service, or organization in order to appear more sustainable than is accurate. Greenwashing can involve vague language, selective disclosure, misleading imagery, or unverified certifications, and it undermines consumer trust and the effectiveness of sustainability initiatives.

**ISO 14001:** An international standard for environmental management systems that guides organizations in systematically improving environmental performance, complying with regulations, and reducing negative environmental impacts through continuous improvement.

**ISO 26000:** An international guidance standard that helps organizations understand and implement social responsibility, covering ethical behavior, human rights, labor practices, environmental impact, and community involvement, developed through a multi-stakeholder process.

**Key Performance Indicators (KPIs):** Quantifiable metrics used to assess progress toward specific goals. In sustainability, KPIs can cover environmental (e.g., greenhouse gas emissions, energy efficiency), social (e.g., workforce diversity, community investment), and economic (e.g., cost savings, revenue from sustainable products) dimensions.

**Life Cycle Analysis (LCA):** A method for assessing the total environmental, social, and economic impacts of a product or service across its entire life cycle, from raw material extraction through production, use, and disposal. LCA informs policy, business decisions, and sustainable design by highlighting areas for improvement and reducing unintended consequences.

**Linkages:** Connections between the three pillars of sustainability (economic, socio-cultural, and environmental) that show how progress in one area can support progress in the others. Creating linkages means developing strategies where the pillars work together rather than separately.

**Millennium Development Goals (MDGs):** A set of 8 global targets used from 2000 to 2015, created by the United Nations. They focused mainly on reducing extreme poverty, improving health, and increasing access to education. They served as an early framework for measuring sustainability-related progress.

**Paris Agreement:** A global climate accord adopted in 2015 in which nearly every nation committed to limiting global temperature rise to well below 2 °C (3.6 °F), with an aspiration of 1.5 °C (2.7 °F), above pre-industrial levels. The agreement establishes shared climate responsibilities while recognizing differences in capacity between nations and includes pledges for climate finance, emissions reductions, and long-term cooperation.

**Perceived Behavioral Control:** An individual's perception of their ability to perform a behavior, considering available resources, skills, and potential obstacles. Greater perceived control increases the likelihood of taking action.

**Productivity:** The capacity to generate more value from the same or fewer inputs through innovation, skills, technology, and improved systems. Productivity drives long-term growth by raising incomes, improving living standards, and enabling economies to use resources more sustainably. It supports development by encouraging smarter ways of working rather than simply increasing effort.

**SMART Goals:** A framework for setting objectives that are Specific, Measurable, Achievable, Relevant, and Time-bound. SMART goals ensure that targets are practical, trackable, and aligned with broader strategic or sustainability objectives.

**Socio Cultural:** This term refers to the combined influence of society and culture on human well-being and development. It focuses on how education, justice, health, identity, and community life shape the strength and resilience of a population. It is the pillar of sustainability that emphasizes empowering people, protecting cultural traditions, and creating inclusive and fair communities.

**Social Justice:** This means ensuring fairness, inclusion, and equal rights for all people. It involves access to legal systems, participation in governance, protection of rights, and support for people who face inequality. In this chapter, social justice includes expanding social protection, creating equal opportunities, protecting cultural identity, and promoting gender equality so that communities can participate fully and thrive.

**Theory of Planned Behavior:** A psychological framework developed by Icek Ajzen in 1991 that explains how individual behavior is guided by three key factors: attitudes toward the behavior, subjective norms, and perceived behavioral control. It is widely used to understand and promote sustainable behaviors.

**Tradeoffs:** Situations in sustainability where improving one or two pillars (economic, socio-cultural, environmental) creates negative consequences for another. For example, a project might create jobs (economic benefit) and improve education (social benefit) but require removing trees or altering land (environmental cost).

Tradeoffs often cannot be avoided, so decision makers must evaluate whether the benefits outweigh the drawbacks.

**United Nations:** An international organization founded in 1945 and composed of 193 Member States. The United Nations promotes global peace, security, human rights, development, and international cooperation. It established the Sustainable Development Goals as part of its 2030 Agenda for Sustainable Development.

**United Nations Global Compact:** A voluntary international initiative encouraging companies to adopt principles in human rights, labor standards, environmental protection, and anti-corruption, providing a global framework for implementing CSR.

**Subjective Norms:** Social pressures or expectations perceived by an individual, including the approval or disapproval of important people or social groups, which influence whether they engage in a behavior.

**Sustainable Development Goals (SDGs):** A set of 17 global goals set by the United Nations and adopted in 2015 by nearly every country. The goals aim to be achieved by 2030 and cover a wide range of sustainability issues such as ending hunger, reducing inequality, addressing climate change, and promoting peace and justice. They expand on the MDGs by providing clearer, measurable sustainability targets.

## 1.2 Appendix B. Commonly Used Abbreviations for Organizations

**AFDB (African Development Bank):** A regional multilateral development bank that promotes sustainable economic development and social progress across African nations, often supporting projects related to climate resilience, infrastructure, and poverty reduction.

**ASEAN (Association of Southeast Asian Nations):** A regional intergovernmental organization that promotes economic cooperation, political stability, and sustainable development among Southeast Asian countries.

**C40 (C40 Cities Climate Leadership Group):** A global network of major cities working together to reduce greenhouse gas emissions and build climate resilience through coordinated urban sustainability initiatives.

**CDC (Centers for Disease Control and Prevention):** A national public health agency that supports disease prevention, health monitoring, and emergency preparedness, often referenced in global health and resilience discussions.

**CDP (Carbon Disclosure Project):** An international nonprofit organization that runs a global disclosure system for companies, cities, states, and regions to measure and manage their environmental impacts. CDP collects data on greenhouse gas emissions, climate risks, water management, and deforestation, providing transparency for investors, policymakers, and the public.

**EC (European Commission):** The executive branch of the European Union responsible for proposing legislation, implementing decisions, and managing EU sustainability, climate, and economic policies.

**EU (European Union):** A political and economic union of member states that develops environmental regulations, climate strategies, and sustainability frameworks applied across Europe.

**FAO (Food and Agriculture Organization of the United Nations):** A global agency that leads international efforts to end hunger, improve nutrition, and promote sustainable agriculture through research, policy guidance, and food-system monitoring.

**G7 (Group of Seven):** A coalition of seven major industrialized countries that addresses global economic governance, sustainability challenges, and climate policy coordination.

**G20 (Group of Twenty):** An international forum of major economies that works on global financial stability, inclusive growth, and climate-related economic strategies.

**GRI (Global Reporting Initiative):** An international organization that sets widely used sustainability reporting standards, helping companies measure and communicate environmental, social, and governance impacts.

**IADB (Inter-American Development Bank):** A regional development bank that supports sustainable economic development, climate adaptation, and infrastructure investment across Latin America and the Caribbean.

**IEA (International Energy Agency):** A global energy authority that provides data, policy guidance, and analysis on energy efficiency, renewable energy, and climate mitigation strategies.

**ILO (International Labour Organization):** A United Nations agency that promotes labor rights, good working conditions, and inclusive economic development.

**IMF (International Monetary Fund):** A global financial institution that supports economic stability, development financing, and policy guidance, including sustainability-linked macroeconomic issues.

**IPBES (Intergovernmental Science-Policy Platform on Biodiversity and Ecosystem Services)**: A global intergovernmental body that assesses the state of biodiversity and ecosystem services, providing policymakers with science-based information to support conservation, sustainable use of natural resources, and policy decisions that address biodiversity loss and ecosystem degradation.

**IPCC (Intergovernmental Panel on Climate Change):** A scientific body that assesses climate change research to inform global climate policy, adaptation strategies, and sustainability planning.

**IRENA (International Renewable Energy Agency)**: A global intergovernmental agency that drives adoption and sustainable use of renewable energy, including bioenergy, geothermal, hydropower, ocean, solar and wind energy.

**ISO (International Organization for Standardization):** A international nongovernmental organization that brings global experts together to agree upon standard and managing process in order to improve safety.

**IUCN (International Union for Conservation of Nature):** A global organization focused on nature conservation, biodiversity protection, and sustainable resource management.

**NDC (Nationally Determined Contribution):** A commitment submitted by each country under the Paris Agreement that outlines national plans for reducing emissions and adapting to climate change.

**OECD (Organisation for Economic Co-operation and Development):** An international organization that supports evidence-based policy development on economic resilience, sustainability, and social well-being.

**REDD+ (Reducing Emissions from Deforestation and Forest Degradation):** A United Nations-backed program that provides financial incentives to developing countries to reduce greenhouse gas emissions from deforestation and forest degradation, conserve forest carbon stocks, and sustainably manage forests. REDD+ is designed to support climate mitigation and biodiversity protection while promoting sustainable development.

**UN (United Nations):** An international organization composed of member states that leads global efforts on peace, development, human rights, sustainability, and climate action.

**UNCTAD (United Nations Conference on Trade and Development):** A UN body that supports developing countries in global trade, sustainable development, and economic diversification.

**UNDP (United Nations Development Programme):** A UN agency that helps countries address poverty, inequality, governance, and climate resilience through policy guidance and development financing.

**UNDRR (United Nations Office for Disaster Risk Reduction):** A UN agency that coordinates global disaster risk reduction efforts and helps countries build climate and hazard resilience.

**UNDESA (United Nations Department of Economic and Social Affairs):** A UN agency that brings the global community together to work towards common solutions to the world's most pressing problems by helping countries translate their global commitments into national action in the economic, social and environmental spheres.

**UNECE (United Nations Economic Commission for Europe):** A regional UN body supporting sustainable development, trade, clean energy, and environmental cooperation across Europe and beyond.

**UNEP (United Nations Environment Programme):** The leading global environmental authority that advances sustainability through climate policy, environmental monitoring, and ecosystem protection.

**UNESCO (United Nations Educational, Scientific and Cultural Organization):** A global organization that builds peace through cooperation in education, science, culture, and communication, and protects both tangible and intangible cultural heritage.

**UNFCCC (United Nations Framework Convention on Climate Change):** An international treaty responsible for coordinating global efforts to combat climate change, including oversight of the Paris Agreement.

**UNHCR (United Nations High Commissioner for Refugees):** A global organization that protects refugees and displaced persons, addressing humanitarian challenges linked to conflict, climate, and environmental change.

**UNICEF (United Nations Children's Fund):** A UN agency that supports children's health, education, safety, and rights, especially in vulnerable communities.

**UNIDO (United Nations Industrial Development Organization):** A UN body focused on promoting inclusive and sustainable industrial development, green technologies, and clean production.

**UNWTO (United Nations World Tourism Organization):** A UN agency that promotes sustainable tourism as a driver of economic growth, cultural exchange, and environmental protection.

**WHO (World Health Organization):** A United Nations agency that leads global public health initiatives, disease prevention, and health system strengthening to support community resilience and sustainable development.

## 1.3 Appendix C. Resources for Diving Deeper

**Social Impact Tools**

1. Social Return on Investment (SROI)

**Key resource**

Social Value International provides the official SROI framework, methodological guidance, and standards at https://www.socialvalueint.org/guide-to-sroi

**What the tool provides**

Social Return on Investment is a framework that assigns monetary values to social, environmental, and economic outcomes that are not typically captured in financial accounting. It combines stakeholder engagement, outcome mapping, and valuation techniques to translate social change into comparable financial terms.

**Purpose of the tool**

The purpose of SROI is to assess whether the social value created by a program, policy, or investment outweighs the resources invested. It is particularly useful when decision makers need to compare alternative interventions or justify funding based on impact rather than outputs alone.

**Potential outcomes of using the tool**

Using SROI results in a ratio that expresses social value created per unit of investment, such as three dollars of social value generated for every one dollar invested. Additional outcomes include clearer identification of high impact activities, improved accountability to stakeholders, and stronger evidence for funding and policy decisions.

2. Gini Coefficient and Income Inequality Metrics

**Key resource**

The World Bank's Gini Index and related inequality data are available through its open data portal at https://data.worldbank.org/indicator/SI.POV.GINI

**What the tool provides**

The Gini Coefficient is a statistical measure of income or wealth distribution within a population. It is often presented alongside Lorenz curves and related inequality indicators that allow comparison across countries, regions, and time periods.

**Purpose of the tool**

The primary purpose of the Gini Coefficient is to assess the degree of inequality within a society. It supports analysis of social equity, distributive justice, and the social sustainability implications of economic growth.

**Potential outcomes of using the tool**

Using Gini data allows researchers and policymakers to identify rising or declining inequality, compare inequality levels across contexts, and assess whether development outcomes are broadly shared or concentrated among certain groups.

## Poverty and Human Development Indices (MPI and HDI)

**Key resources**

UNDP Human Development and composite index data are available at https://hdr.undp.org/data-center/composite-indices

The Oxford Poverty and Human Development Initiative provides detailed MPI methodology and data at https://ophi.org.uk/multidimensional-poverty-index

**What the tools provide**

The Human Development Index combines indicators of life expectancy, education, and income into a single composite measure of overall development. The Multidimensional Poverty Index measures deprivation across health, education, and living standards at the household level, rather than relying solely on income.

**Purpose of the tools**

These indices are designed to move beyond purely economic measures of progress and capture broader aspects of human wellbeing. They are widely used to monitor development outcomes, assess social conditions, and inform international and national policy.

**Potential outcomes of using the tools**

Using HDI enables comparison of overall human development across countries and over time. Using MPI helps identify who is poor, how they are poor, and which deprivations matter most. Together, these tools support targeted interventions, social policy design, and evaluation of progress toward the Sustainable Development Goals.

4. Gender Inequality Index (GII)

**Key resource**
Official Gender Inequality Index data and documentation are provided by UNDP at https://hdr.undp.org/en/composite/GII

**What the tool provides**
The Gender Inequality Index measures disparities between women and men across reproductive health, empowerment, and labor market participation. It combines indicators such as maternal mortality, educational attainment, political representation, and workforce participation into a single score.

**Purpose of the tool**
The purpose of the GII is to assess structural gender inequalities that limit social inclusion, economic participation, and human development. It provides a standardized way to compare gender gaps across countries and over time.

**Potential outcomes of using the tool**

Using the GII highlights areas where gender disparities are most pronounced, supports gender responsive policy development, and enables monitoring of progress toward gender equality goals. It is also useful for evaluating social sustainability and equity dimensions within development and ESG frameworks.

5. World Bank Poverty and Inequality Platform (PIP)

**Key resource**
The Poverty and Inequality Platform is accessible at https://pip.worldbank.org

**What the tool provides**
The Poverty and Inequality Platform is an interactive data system that integrates poverty, inequality, and shared prosperity indicators. It allows users to explore trends, compare countries, and download datasets.

**Purpose of the tool**
The purpose of PIP is to provide transparent, comparable, and regularly updated data for monitoring global and national poverty and inequality trends.

**Potential outcomes of using the tool**

Using PIP supports evidence based policy analysis, academic research, and evaluation of social development outcomes. It enables users to track progress, identify vulnerable populations, and assess distributional impacts of growth.

6. OECD Wellbeing and Social Indicators Dashboards

**Key resource**
OECD dashboards and tools are available at https://www.oecd.org/en/data/dashboards-and-tools.html

**What the tools provide**
OECD dashboards offer multidimensional indicators on wellbeing, inequality, social protection, trust, and quality of life. They include interactive visualizations and downloadable data for member and partner countries.

**Purpose of the tools**
These dashboards are designed to complement economic indicators by providing a broader view of societal progress and social sustainability.

**Potential outcomes of using the tools**

Using OECD social dashboards enables comparative analysis, policy benchmarking, and integration of wellbeing indicators into national and regional decision making.

## Economic Measurement Tools

7. Key Performance Indicators (KPIs) for Sustainability

**Key resource**

The Global Reporting Initiative provides standardized sustainability indicators and reporting guidance at https://www.globalreporting.org/standards

**What the tool provides**

Sustainability focused Key Performance Indicators are quantitative measures used to track organizational performance across economic, environmental, and social dimensions. Examples include revenue from sustainable products, energy intensity, emissions per unit of output, and workforce composition indicators.

**Purpose of the tool**

The purpose of sustainability KPIs is to translate strategic sustainability goals into measurable, trackable metrics. They support internal management, external reporting, and alignment with international reporting standards.

**Potential outcomes of using the tool**

Using KPIs enables organizations to monitor progress over time, benchmark performance against peers, identify efficiency gains, and communicate sustainability performance to stakeholders in a consistent and transparent manner.

8. Eco Efficiency Metrics

**Key resource**

The World Business Council for Sustainable Development provides guidance on eco efficiency concepts and metrics at https://www.wbcsd.org/programs/climate-and-energy/energy/eco-efficiency

**What the tool provides**

Eco efficiency metrics link economic value creation with environmental impact by expressing performance as a ratio, such as value added per unit of energy use or emissions. These metrics integrate productivity and environmental performance into a single indicator.

**Purpose of the tool**

The purpose of eco efficiency metrics is to support decision making that improves environmental performance while maintaining or enhancing economic output.

**Potential outcomes of using the tool**

Using eco efficiency metrics helps organizations identify opportunities to reduce resource use, lower emissions, and improve cost effectiveness while maintaining competitiveness.

9. Life Cycle Costing (LCC)

**Key resource**

The UNEP Life Cycle Initiative provides methodological guidance and tools for life cycle costing at https://www.lifecycleinitiative.org

**What the tool provides**

Life Cycle Costing evaluates the total cost of a product, service, or project across all stages of its life cycle, including acquisition, operation, maintenance, and end of life. It can incorporate environmental and social costs alongside direct financial costs.

**Purpose of the tool**

The purpose of LCC is to support long term decision making by revealing costs that may not be visible in short term budgeting or procurement processes.

**Potential outcomes of using the tool**

Using LCC supports more informed investment decisions, identification of cost effective sustainability solutions, and integration of environmental considerations into economic planning.

**Environmental Measurement Tools**

10. Carbon Footprint Calculators

**Key resources**

Carbon Footprint Ltd provides organizational and product level emissions calculators at https://www.carbonfootprint.com/calculator.aspx

The United States Environmental Protection Agency offers a public carbon footprint calculator at https://www.epa.gov/carbon-footprint-calculator

**What the tools provide**

Carbon footprint calculators estimate greenhouse gas emissions associated with activities, products, or organizations. Results are typically expressed in tons of carbon dioxide equivalent and broken down by emissions source.

**Purpose of the tools**

The purpose of carbon footprint calculators is to quantify emissions in order to support measurement, reporting, and reduction planning.

**Potential outcomes of using the tools**

Using these tools enables organizations and individuals to identify high emission activities, track emissions reduction progress, and support climate related targets and disclosures.

11. Water Footprint and Water Risk Tools

**Key resources**

The Water Footprint Network provides methodology and tools for water footprint assessment at https://waterfootprint.org

**What the tools provide**

Water footprint tools measure direct and indirect water use associated with products, activities, or supply chains. Water risk tools assess exposure to water scarcity, pollution, and regulatory risks.

**Purpose of the tools**

The purpose of these tools is to identify water intensive processes and assess sustainability risks related to water availability and quality.

**Potential outcomes of using the tools**

Using water footprint and risk tools supports prioritization of water efficiency measures, improved water stewardship, and more resilient supply chain planning.

12. Life Cycle Assessment (LCA) Tools

**Key resources**

OpenLCA provides open source software for life cycle assessment at https://www.openlca.org

**What the tools provide**

Life Cycle Assessment tools quantify environmental impacts associated with a product or service across its entire life cycle, including emissions, energy use, water consumption, and waste generation.

**Purpose of the tools**

The purpose of LCA is to support evidence based comparison of products, processes, or design alternatives based on environmental performance.

**Potential outcomes of using the tools**

Using LCA tools enables improved product design, reduced environmental impacts, and more credible environmental claims and disclosures.

## 1.4 Appendix D. Student Guide: Exploring SDG Data Further

1. UN SDG Global Database

**Website:** https://unstats.un.org/sdgs/
**What you can do:**

- Access official SDG indicators for all countries.
- Use dashboards to visualize trends over time.
- Download datasets in Excel or CSV format.
- Explore disaggregated data by sex, age, income, and region.

**Tips:**

- Use the "Indicators" tab to quickly find specific SDGs.
- Compare multiple countries or regions for your research.
- Check the "Data Explorer" to create charts and maps.
    - Visualize SDG indicators across countries and regions on an interactive world map.
    - Compare progress for multiple countries at a glance.

- Filter data by year, indicator, or region to focus on specific areas of interest.
- Animate maps over time to observe trends and improvements in SDG performance.
- Customize color schemes to highlight patterns, disparities, or progress levels.
- Export map images for presentations or reports to support your analysis.

World Bank World Development Indicators

**Website:** https://data.worldbank.org

**What you can do:**

- Access over 1,000 indicators covering economic, social, and environmental dimensions.
- Identify indicators that map directly to SDG targets (poverty, education, energy, health, water).
- Download datasets for detailed analysis.
- Use the built-in visualization tools or the API for advanced data work.

**Tips:**

- Start with the "Explore Data" search bar to find relevant indicators.
- Export data to Excel, Google Sheets, or statistical software.
- Combine World Bank data with UN SDG indicators to get a complete picture.
- Use trend analysis to see changes over time.
- Explore the interactive maps: Many WDI indicators are visualized through world maps, which allow you to:
  - Compare countries or regions at a glance for a single year or over time.
  - Identify geographic patterns, such as regions with higher poverty or lower access to clean energy.
  - Use filters to focus on specific income groups, regions, or countries.
  - Animate maps over time to observe trends and shifts globally.
  - Customize map views by selecting color scales, thresholds, or clustering to highlight significant differences.
  - Combine map views with charts to provide both spatial and temporal perspectives for research.

**Practical Research Advice for Students**

To make the most of these resources:

1. **Start with a clear research question**: Decide which SDG or indicator you want to explore before diving into the data.
2. **Use dashboards for visual exploration**: Both the UN and World Bank sites allow you to generate charts and maps quickly to identify trends or patterns.
3. **Download datasets for detailed analysis**: Export data to Excel, Google Sheets, R, or Python to perform calculations, create graphs, or conduct statistical tests.
4. **Disaggregate data**: Look at breakdowns by gender, age, region, or income to understand differences within countries.

5. **Compare multiple countries or regions**: Use side-by-side analysis to identify which countries are progressing faster or slower and hypothesize why.
6. **Combine datasets**: Use both UN SDG data and World Bank indicators to create a more complete picture. For example, link economic indicators from the World bank with social indicators from the UN database.
7. **Check data frequency and coverage**: Some indicators are reported annually, others less often. Make sure the timeframes match your research needs.
8. **Document your sources**: Always note the exact URL and year of access so your work can be verified and updated later.

The manufacturer's authorised representative in the EU is Springer Nature Customer Service Centre GmbH, Europaplatz 3, 69115 Heidelberg, Germany. If you have any concerns regarding our products, please contact ProductSafety@springernature.com

Printed and bound by CPI Group (UK) Ltd, Croydon, CR0 4YY

24/06/2026

02147159-0001